energy independence & industrial renaissance

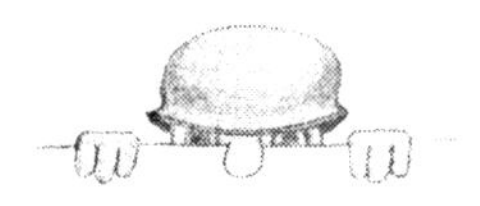

JAMES B. EDWARDS

Aventine Press

Cover photo courtesy of NASA.

First Edition

Published by Aventine Press
750 State St. #319
San Diego CA, 92101
www.aventinepress.com

ISBN: 1-59330-642-3

Library of Congress Control Number: 2007941152
Library of Congress Cataloging-in-Publication Data
There Is A Silver Bullet II

Printed in the United States of America

INTRODUCTION

Once upon a time, America was the industrial leader and envy of the world. Today, China and India are pursuing massive crash programs designed to seize that position. Fantastic new technologies present America with the opportunity to recapture its title role. America must reindustrialize. Energy is the biggest business in the world; electricity is the energy of the future. Nikola Tesla's dream will be realized; the world is on the threshold of a new age of electricity. There is an arsenal of new technology magic bullets primed to give America total energy independence and economic revitalization. Reindustrialization and massive electrification will launch a new era of economic prosperity. America consumes 140 billion gallons of gasoline per year. America imports 70% of its oil at a cost of $700 billion per year - the greatest transfer of wealth in the history of the world. America can achieve total energy independence and economic rebirth by pursuing the following technologies: PHEVs, superbatteries/supercapacitors, algae bio-fuel, uncap old oil wells, switchgrass, oil shale, natural gas, nuclear power, solar, wind, geothermal, nanotechnology, robotics, etc, etc.

In the past 150 years (since 1859), the entire world has consumed one trillion barrels of oil.
America has two trillion barrels of residual oil in 400,000 capped-off oil wells.
Beneath the Rocky Mountains, in the vast Bakken Formation, America has two trillion barrels of light, sweet crude oil.
Beneath the Appalachian Mountains, America has 2000 trillion cubic feet of natural gas; enough to last 100 years.

The Canadian tar sands contain over two trillion barrels of oil.
America has 275 billion tons of coal, enough to last 300 years; convertible to trillions of barrels of diesel oil.
There are over 1700 1000 megawatt coal-fired power plants in the U.S. that could produce 170 billion gallons of bio-fuel per year if equipped with algae bio-reactors to capture the CO_2 emissions.
Fifty million acres of switchgrass (the area of South Dakota) could produce over 140 billion gallons of ethanol per year.
Battery/capacitor energy densities of over 1000 watt hours per kilogram will bring electric cars able to travel 500 miles on a single charge; eliminating the need for engines or fuel.

Quick fixes abound:
Andy Grove plans to convert 10 million PSVs to PHEVs in 4 years.
T. Boone Pickins plans to convert all our diesel-powered 18-wheelers to CNG in 5-7 years.
Nanotechnology is projected to become a $3 trillion industry by 2015.
Robotics could become a $5 trillion to $10 trillion industry in ten years.

PHEVs: America consumes 140 billion gallons of gasoline per year. Plug-in electric cars can make America energy independent. If all American cars were plug-ins with an electric only range of 40 miles, America would not have to import one single gallon of petroleum from the Middle East. DOE secretary Steven Chu wants to direct most of its resources to building electric vehicles. The average American drives about 25 to 30 miles per day. The average European drives about 17 miles per day. Plug-ins like the GM Volt, with an electric only range of 40 miles, can reduce U.S. gasoline consumption by 85%. European car makers are designing cars to have an electric only range of 100 kilometers – 62 miles; with these vehicles fuel will be required only for long trips. Replacing America's 250 million cars with PHEVs will take over ten years. Andy Grove, former Chairman and CEO of Intel has a plan to beat this problem. There are about 80 million gas-guzzling Pickups, SUVs, and Vans (PSVs) on the roads. Grove says 10 million of these PSVs can be converted to PHEVs in four years. PSVs have plenty of available space to be easily converted to PHEVs. Even if the total electric driving range is no more than 50 miles between recharges, converting these 80 million PSVs to PHEVs would cut petroleum imports by 50% to 60%. Grove expects battery prices to drop 50% in five years. Grove calls for a 50% tax credit for retrofitting the PSVs, and free electricity for two years. Millions of new jobs would be created. Extending Grove's plan to convert the other 70 million PSVs to PHEVs would accelerate the process. Grove's plan looks like a good way to get started. Researchers at Stanford University have discovered a way to use silicon nanowires to give lithium-ion (Li-ion) batteries as much as ten times more energy per charge. Meaning a PHEV with an electric range of 40 miles (like GM's Volt) would have a range of 400 miles, enabling it to have a much smaller engine or none at all. IBM's "Battery 500" project hopes within two years to develop a lithium-air battery that would have ten times the energy density of current lithium-ion batteries; giving cars an electric range of at least 500 miles between charges. Felix Kramer wants to take Andy Grove's plan one step further by converting, retrofitting, or fixing up America's 250 million vehicle fleet. He wants to adopt the software industry model where more money is made from upgrades than from original sales; he envisions hundreds of thousands of new jobs. Retrofitting is SOP in DOD, where equipment must last for decades. B-52s have been flying for 50 years and may fly for 50 more. Right now, the Abrams tank and

Bradley fighting vehicle inventories are being updated for the long term. Once low-cost superbatteries/ supercapacitors are in mass production, the resulting simplified running gear will open the car business to thousands of small entrepreneurs. America will return to the pioneer days of the car industry when there were over 2000 car manufacturers; that should make the timing about right for the Return of the *Roaring 20's! 23 Skidoo!*

SUPERCAPACITORS: Every now and then a new technology comes along that changes civilization, think of gunpowder, steam engines, electricity, the internal combustion engine, nuclear energy, transistors, integrated computer circuits, the internet. We may be on the threshold of another such disruptive technology – supercapacitors. Should this technology become viable, the world energy system would be totally transformed. A cheap and reliable means of storing electricity would make the wind and solar industries independent of intermittent energy sources. Supercapacitors installed along the national smart grid would enable enormous gains in distribution efficiency. Small inexpensive home storage units would make rooftop solar and wind competitive with centralized power generation and solve the emergency power problem. If low cost cars and trucks can be built that have a 250 mile all electric range and are instantly rechargeable, the automotive and petroleum industries would be transformed forever.

Supercapacitors could launch the world into the age 100% electric vehicles. EEStor Inc of Cedar Park, Texas claims to have designed a supercapacitor that can be recharged in five minutes and give a car a range of 500 miles per charge. DOD mega-contractor Lockheed-Martin bought exclusive rights to EEStor's power system for military applications. Lockheed-Martin said EEStor's technology "could lead to energy independence for the warfighter" L-M said EEStor's supercapacitor holds 10 times the energy at 1/10th of the weight of lead-acid batteries. Not only every vehicle, but every home and office building in America, will have a supercapacitor auxiliary power system. Supercapacitors will be the “holy grail” of a new age of electricity – alone, they may solve the world's energy problem!

DOE has said the greatest impediment to the growth of the wind and solar industries is energy storage. Storage is vital for times when the wind is not blowing and the sun is not shining. Some of the advantages of supercapacitors over storage technologies like batteries include: higher power capacity, longer life, wider thermal operating range, lighter weight, and lower maintenance.

MIT's Technology Review reported EEStor's ambitious goal is nothing less than to “replace the electrochemical battery” in almost every application from hybrid-electric and pure-electric vehicles to laptop computers to utility-scale electrical storage. EEStor says their cell phone EESUs will store three to five times more energy than existing lithium batteries, never degrade, and be rechargeable in seconds. By utilizing EESU for grid-load-leveling, 45% more electricity can be put on the grid to supply the electricity needs of the electric vehicle market as it emerges. EESUs will make the wind and solar industries viable. Wind farms and solar arrays will be able to operate like coal-fired power plants.

The EESU is a solid-state battery, its ceramic ultracapacitor has a barium-titanate dielectric (insulator) that achieves extremely high specific energy – the amount of energy in a given unit of mass. EEStor claims a specific energy of about 280 watt hours per kilogram, compared with about 120 watt hours per kilogram for lithium-ion batteries and 32 watt hours per kilogram for lead-acid batteries. Fluidic Energy, a spin-off of Arizona State University, recently received an ARPA-E award from DOE to develop metal-air battery technology. Fluidic believes metal-air batteries are potentially 11 times more energy dense than lithium-ion batteries, and can be built for half or even a third of the cost. Fluidic hopes to achieve energy densities of from 900 to 1600 watt hours per kilogram; that could lead to electric vehicles that could travel 400 to 500 miles on a single charge. This opens up a galaxy of applications ranging from pacemakers to laptops to electric vehicles to locomotives to electric utilities to military applications like catapults, directed-energy weapons and hyper-velocity cannons.

HYPERCARS: IBM's "Battery 500" project is developing lithium-air batteries that will have ten times the energy density of current lithium-ion batteries. Stanford University researchers are developing silicon nanowires that will give batteries ten times more energy density. Arizona State University spin-off Fluidic Energy is developing metal-air batteries that will have eleven times the energy density of lithium-ion batteries. Such superbatteries will eliminate the need for engines, transmissions, driveshafts, universal joints, gastanks, and perhaps axles.

Ultralight and ultra-low-drag platforms yield enormous advantages in mass decompounding. Carbon-fiber composites can absorb ten times more crash energy than steel. A composite structure with the same strength as a steel structure will be two to three times lighter. Nanotubes are roughly 30 to 100 times stronger than steel and 10 to 100 times stronger than carbon-fiber. Hypercars will be built like Indy 500 racecars. Millions have seen Indy 500 cars crash into the walls at 200mph and watched the driver get out and walk away. Rocky Mountain Institute (RMI)said the strength of a carbon-fiber monocoque structure can be demonstrated by watching someone trying to eat an Atlantic lobster without tools.

Hypercars will have 2 to 3 times lower curb mass, 2 to 6 times lower aerodynamic drag, 3 to 5 times lower rolling resistance than today's conventional car. Today's average production sedan has a drag coefficient (Cd) of 0.33 and a curb weight of 3500-4000 pounds. A Hypercar will have a Cd under 0.20 and a curb weight of around 1500 pounds. Superbatteries will give the Hypercar a range of over 1000 miles on a single charge.

Mechanical systems will be replaced by drive-by-wire systems: brake-by-wire, steer-by-wire, suspension-by-wire, etc. All drive-by-wire systems will be redundant as on aircraft. All glazing will be made of new polycarbonates, similar to those used in headlight covers; which are 50% lighter than glass. To minimize unsprung mass, the two front electric motors and brakes will be mounted inboard, connected to the wheels by carbon-fiber half-shafts; the two rear motors will be mounted within the wheel hubs; this produces a lower rear floor and better underbody aerodynamics. It will be equipped with run-flat wheels and tires that can be driven at highway speeds on four flat tires. It will be equipped with a sidestick control system for controlling steering, braking, and acceleration. Sidesticks get rid of the steering column and pedals; the leading cause of injuries in collisions, especially for women drivers. Sidestick control permits much faster reaction times and evasive maneuvers; all high-performance aircraft utilize sidesticks. The Hypercar will not be a car with chips, but a computer with wheels. The auto industry of the future will be run from the Silicon Valleys of the world instead of the Detroits. By eliminating so many complex and expensive components, the cost an complexity of manufacturing automobiles will be enormously reduced. Low volume construction technology, as espoused by RMI, will permit thousands of small "skunk works" manufacturers to emerge. In the early 1900s, there were over 2000 car manufacturers in the U.S. We should see a rebirth of this “golden era.”

ALGAE: An acre of corn produces approximately 400 gallons of ethanol per year; an acre of switchgrass produces 1500 gallons per year; an acre of sugarcane produces 3000 gallons per year; an acre of algae produces up to 100,000 gallons of bio-fuel per year. About 150 million acres of corn would be required to produce enough ethanol annually to replace gasoline. That’s an area about the size of California and South Dakota combined. About 50 million acres of switchgrass could produce enough ethanol annually to replace gasoline. That's an area about the size of South Dakota. Algae at a production rate of 15,000 gallons/acre/year, would require about 9.3 million acres to produce 140 billion gallons of biodiesel and ethanol. Vertigro Energy of El Paso, Texas claims to have a process that produces 100,000 gallons/acre/year. At that rate, about 1.4 million acres (Rhode Island has 776,957 acres) could produce enough fuel to meet all of America’s transportation needs. Many companies are developing new algae production technologies. One company claims to have a process that produces 180,000 gallons/acre/year. When the sun shines, plants break down water into hydrogen and oxygen, using these to turn CO_2 into glucose, and venting off oxygen as a waste product. All plant life lives on CO_2, it is their food. During photosynthesis, plants use sunlight to recycle CO_2 into fuel. One plant is very good at this process – algae. Algae is a tiny single-cell plant that produces most of the Earth’s atmosphere. Algae recycles CO_2 into fuel, the more CO_2, the more algae. Algae bioreactors use CO_2 piped from the smokestacks of powerplants to greatly accelerate this process.

The process works equally well for any plant that emits large amounts of CO_2 – coal-fired powerplants, steel plants, aluminum plants, wastewater treatment plants, breweries, etc., etc. We can almost completely eliminate industrial CO_2 emissions by using algae to convert them to biofuels. Green Fuel Technologies, founded by MIT engineer Cary Bullock, says one 1000 megawatt coal-fired powerplant can produce over 100 million gallons of biodiesel and ethanol per year (about a 50/50 split). Such a plant requires an algae farm of about 2000 acres. CEO Bullock says: "All we're doing is what Mother Nature does, only we do it faster." Thus, about 1500 1000 megawatt plants could produce about 140 billion gallons of biofuel per year. Nationwide, there are about 1000 coal-fired powerplants with enough space nearby for a few hundred to a few thousand acres of algae farm. There are 1750 coal-fired powerplants in the U.S. that have space for 250 acre algae farms. Algae bioreactors fed by CO_2 producing industrial plants could convert all the world's industrial nations from fuel importors to fuel exportors.

One 1000 megawatt coal-fired power plant equipped with an algae bio-reactor to capture its CO_2 emissions, can produce 100 million gallons of bio-fuel per year. There are over 1700 1000 megawatt power plants in the U.S. that together could produce 170 billion gallons of biofuel per year. More than enough to offset the approximately 140 billion gallons of gasoline we consume per year. There are over 300,000 factories in the U.S., most of them powered by coal. Equipping these with algae bio-reactors would produce hundreds of billions of gallons of biofuel per year in addition to the 170 billion gallons per year produced by the larger coal-fired power plants. America would be awash in an ocean of bio-fuel! Sale of the biofuel could offset the cost of the coal; making the electricity free, or it could fund the national health care plan. America has over 275 billion tons of coal; enough to last 300 years at the current rate of usage.

Exxon and Synthetic Genomics have teamed-up in a $600 million deal to brew algae biofuels in existing refineries. The companies plan to spend the next five or six years trying to find the answers to three problems: what is the most suitable strain of algae; what is the best way to grow it; how do you mass produce it economically using the existing energy infrastructure? Exxon/SGI chose algae after an extensive search for the most suitable biofuel. Algae enjoys many advantages: it does not require arable land or fresh water; it consumes CO_2; it can be fueled by capturing the CO_2 of high emitters like coal-fired power plants and cement factories. Exxon/SGI are taking a different approach to the problem of turning algae into biofuel. They are trying to tweak the algae to excrete a "hydrocarbon-like" liquid that can be run through existing refineries. Most other algae-to-oil processes basically squeeze the algae to extract the oil. Exxon/SGI will launch the project with a new research facility in San Diego. They project it will be five to ten years before small-scale plants are up and running.

UNCAP THE OIL WELLS: Exxon-Mobil estimates that one trillion barrels of oil have been extracted globally since oil was first produced commercially in Titusville, PA in 1859. The Department of Energy (DOE) estimates that capped-off oil wells still contain 65% to 70% of residual oil that was too viscous to extract profitably at the time they were capped. That means there are over two trillion barrels of oil remaining in capped-off oil wells globally. Thus, over twice as much oil lies waiting in capped-off wells as has been consumed by the entire world since 1859. That is also twice as much as current estimates of untapped proven reserves - approximately 1.15 trillion barrels. The U.S.

has over 400,000 capped-off oil wells still containing an ocean of oil. There is no shortage of oil! We just need to uncap the wells and get the oil out with new technologies like microwave extraction. Despite this, oil companies are spending billions of dollars dotting our coast lines with complex and expensive offshore oil rigs. As a follow-on to the 1926 oil depletion allowance, the government through the Enhanced Oil Recovery (EOR) credit, grants oil companies handsome tax credits to go after oil that is more expensive to extract, like especially thick residual crude oil. What are they waiting for? Uncap the wells!

Global Research Corp. (GRC) has developed a process that uses microwaves to extract fuel from shale rock, tar sands, tires and plastics, coalfield waste, and "depleted" oil fields. Every material has a frequency that excites its particular molecules best. The magnetron in your kitchen microwave oven is tuned to 2450 MHz, which is the specific frequency for water molecules. GRC has identified microwave frequences (RF) specific to the target substance. GRC uses much higher RF Klystrons, a microwave electron tube with velocity modulation that is different from magnetrons. GRC's technology will extract oil from shale, tar sands, waste oil, and "depleted" oil fields. GRC estimates there are more than two trillion barrels of oil remaining in capped-off oil wells.

GRC estimates the U.S. has two trillion barrels of oil in oil shale. Worldwide, oil shale holds an estimated 14 trillion barrels of oil; a 500 year supply at current rates of usage. There are over two trillion barrels of oil in Canadian tar sands. Refinery waste oil (a.k.a. heavy oil or slurry oil) is oil that is too thick to economically crack or reprocess with current technology. About 3% to 7% of oil refinery production is slurry oil. About 365 million barrels of slurry oil is produced by U.S. refineries each year. The GRC process can convert slurry oil to natural gas and oil. The GRC system operates in a vacuum; nothing is released into the atmosphere. To lend perspective to the above statistics, the five largest known foreign oil reserves in Saudi Arabia, Canada, Iran, Iraq, and Kuwait, contain an estimated 793 billion barrels.

Hidden 1000 feet beneath the Rocky Mountains lies the largest untapped oil reserves in the world - the Bakken Formation. The Bakken Formation stretches from Northern Montana, through North Dakota and into Canada. The U.S. Geological Service estimates that it contains between 500 billion and two trillion barrels of light, sweet crude oil. That is enough petroleum to fully fuel the U.S. economy for over 2000 years. The Western United States has more proven oil reserves than the rest of the world put together:

8 times as much oil as Saudi Arabia.
18 times as much oil as Iraq.
21 times as much oil as Kuwait.
22 times as much oil as Iran.
500 times as much oil as Yemen.

If that doesn't blow your mind, recall once more: Exxon-Mobil estimated that one trillion barrels of oil have been extracted globally since 1859.

NUCLEAR: The nuclear power industry is enjoying a revival all over the world, including the U.S., where almost 30 proposals for nuclear reactors are awaiting approval by the Nuclear Regulatory Commission. Nuclear reactors are highly profitable; Connecticut recently proposed a windfall profit tax on theirs. Britain, France, China, Canada, Russia, Japan, India, and Finland see nuclear energy as an enormous opportunity to reduce carbon emissions and relieve the world's energy problems. China and India each plan to add at least 30 new nuclear reactors in the next 15 years; Britain is planning to add 10. Japan Steel Company is the only steel works that can cast the 42-foot long, egg-shaped reactor vessels at the core of the reactor; they have a four year backlog of orders for the vessels. Nuclear power was once regarded as an American technology. Currently, the U.S. gets 50% of its electricity from coal and 20% from its 104 nuclear reactors. Nuclear proponents would like to reverse these percentages.

The average 1000 megawatt coal-fired powerplant is fueled every five days by a train with 110 coal cars, each loaded with 20 tons of coal. We now burn one billion tons of coal a year, up from 500 million tons in 1976. A 1000 megawatt nuclear reactor is fueled every two years by fuel rods delivered on flatbed trucks. The fuel rods are only mildly radioactive and can be handled with gloves. The rods are loaded into the reactor, where they will remain for six years; only one-third of the rods are replaced at each refueling. The replaced rods are transferred to a storage pool inside the containment structure, where they will remain indefinitely; three feet of water blocks the radiation. The rods look the same coming out as they did going in, except that they are now more highly radioactive. There is no exhaust, no carbon emissions, no air, water, or ground pollution. At Three Mile Island, the only radioactive release was a puff of steam that emitted the same level of radiation as a single chest x-ray. There was more radioactive fallout on Harrisburg from Chernobyl than from Three Mile Island.

A spent fuel rod is 95% U-238, which is harmless; most of the remaining 5% is useful. The useful parts, U-235 and plutonium (a manmade element produced from U-238), can be recycled as fuel. We currently recycle plutonium from Russian nuclear missiles. Only cesium-137 and strontium-90, which have half-lives of 28 years and 30 years, respectively, need to be stored in protective areas like Yucca Mountain. In 1977, President Carter outlawed nuclear recycling out of fear that our plutonium would be stolen to make nuclear bombs. Britain, Canada, France, and Russia all recycle their nuclear fuel. France has produced 80% of its electricity with nuclear power for the last 25 years. All of their high-level nuclear waste is stored in a single room in Le Harve. No bloody Yucca Mountain storage needed.

The Chinese are getting ready to mass produce Pebble Bed Modular Reactors (PBMR). PBMRs are walk-away safe, meltdown-proof nuclear power reactors that can be mass produced in factories in modules that can be shipped anywhere in the world and assembled like Legos. They plan to build them by the hundreds, if not thousands. Instead of the white-hot fuel rods that power conventional reactors, PBMRs are powered by tennis ball sized graphite spheres packed with tiny pellets of uranium dioxide. Instead of the corrosive, radioactive, superhot water of a conventional reactor, the PBMR uses inert helium gas to cool the core and transfer the energy to the generator. No water means there is no danger of venting radioactive steam, there is no huge cooling tower or billion dollar pressure dome to contain a leak. The reactor is loaded with spheres; three quarters are fuel spheres, one quarter are graphite moderator spheres. Fuel spheres are continually added from the top and removed from the bottom. The removed spheres are measured to see if the uranium has been used up. If it has, the sphere is sent to a lead-lined steel bin in the basement; if it has not, it is reloaded to the top of the core. An average sphere passes through the core about ten times before depletion; the graphite spheres are always reused. Unlike current systems, there is no steaming radioactive pool of spent fuel rods. The diamond-like silicon coating on the fuel spheres will keep the radioactive decay particles safely isolated for about one million years.

Currently, the Chinese are operating a pilot plant at Beijing's Institute of Nuclear and Energy Technology (INET), the HTR-10 (high-temperature reactor, 10 megawatt). With PBMR systems there are no chain reaction, no danger of radioactive venting. A meltdown is impossible. Continuous fueling eliminates the need to shut down the reactor every 12-18 months to change the fuel. The use of a closed cycle helium gas turbine and magnetic bearings yields thermal efficiencies of 45% versus 33% for steam systems. Since all components are replaceable modules, maintenance cycles are significantly shorter. PBMR systems operate with little human intervention; the staff are there to monitor rather than operate. Project director Zhang Zouyi says: "In a conventional reactor emergency, you have only seconds to make the right decision, with the HTR-10, it's days, even weeks - as much time as we could ever need to fix a problem." High temperature PBMRs have great potential as hydrogen machines. High temperature electrolysis is the most efficient way to extract hydrogen from water. Sandia National Laboratories believe efficiencies could top 60%. China's mass production of PBMR systems would enable them to leapfrog the world into the hydrogen age.

Thorium is a naturally occurring silvery-white slightly radioactive material. Fueling nuclear reactors with thorium instead of uranium would produce half as much radioactive waste and reduce the supply of weapons-grade plutonium by 80%. Known reserves of thorium are virtually inexhaustible. Thorium is four or five times as abundant as uranium; it's about as common as lead. Brazil and India appear to possess the lion's share of the world's thorium deposits. India has 67% of the global reserves of monozite, the primary thorium ore. Turkey, the U.S., and Norway have huge deposits. Thorium is a thermal breeder that creates new fuel as it breaks down, in theory, able to sustain a high-temperature reaction indefinitely. A liquid fluoride thorium reactor (LFTR or lifter) would be 50% more efficient than today's light-water uranium reactors. The U.S. has over 175,000 tons of thorium. If our existing uranium reactors were converted to LFTRs, thorium could power America for 1000 years. China, India, France, America, and Brazil are experimenting with myriad thorium reactor designs. India is developing a 300 MW prototype of a thorium-based Advanced Heavy Water Reactor (AHWR); it is scheduled to be operational by 2011. Five more reactors of this type are planned. India plans to meet 30% of its electricity needs with thorium-based reactors by 2030.

The U.S. has 104 nuclear reactors built between 1970 and 1999 that produce 20% of our electricity and 70% of our carbon-free electricity. China recently raised its goal of nuclear reactors to 132; Russia plans two new reactors every year until 2030. India has six reactors under construction and ten more planned. Japan has 55 reactors and plans to have 12 more by 2018. France gets 80% of its electricity from nuclear reactors, and has the lowest electricity rates and lowest carbon emissions in Europe. Senator Lamar Alexander believes the U.S. should build 100 new nuclear plants in the next 20 years. This would raise our percentage of nuclear produced electricity to 40%.

COAL: The U.S. has an estimated 275 billion tons of recoverable coal reserves. Utah has an estimated 60 billion tons of low-sulfur, clean-burning coal. The U.S. has enormous quantities of waste coal. For every two tons of coal mined, up to half ends up in the reject pile – waste coal. States like Pennsylvania,

Illinois, Kentucky, Virginia, Wyoming, and West Virginia have vast quantities of waste coal. Pennsylvania has an estimated 260 million tons of waste coal. South African company, Sasol, developed an advanced version of the Fischer-Tropsch process to convert waste coal into diesel oil. Using the Sasol process, about one ton of waste coal (culm) yields about one barrel of diesel oil. A Penn State University study (2006) estimated that by 2015, coal would produce more than $1 trillion annually in GDP, $360 billion in new household income, and seven million new jobs. This study did not consider the increase in GDP and jobs that would result from algae derived biodiesel or waste coal diesel produced by the microwave or Sasol process.

NATURAL GAS: If you missed out on "Cash for Clunkers", despair not, Congress may have a brand "new deal" for you. The NAT GAS Act of 2009 offers up to $12,500 toward the purchase of a new car or truck powered by compressed natural gas (CNG). The Act offers fleet operators up to $64,000 in tax credits, and up to $100,000 to anyone opening a CNG filling station. Right now, the only CNG powered car available to American buyers is the Honda GX, but about a dozen foreign auto makers offer CNG models overseas. The Act calls for the replacement of diesel-powered 18-wheelers and heavy-duty vehicles with CNG-powered vehicles over a five to seven year period. It is claimed that this will save 2.5 million barrels of oil per day – that is more than we import daily from Saudi Arabia and Venezuela combined. CNG boosters claim it is the only fuel that can replace diesel in semis and other heavy-duty vehicles. They say a gallon equivalent of CNG is about half the price of gasoline or diesel, and only produces about one-third of the harmful emissions. Shortages of natural gas have been forecast for ages, but as a result of recent record discoveries, supplies have surged over 50% - America is swimming in the stuff. The Marcellus Shale field runs from West Virginia through Pennsylvania and into New York and has as much natural gas as the North Field in Qatar, the largest field ever discovered. America has 2000 trillion cubic feet of natural gas reserves mostly in Appalachia, Arkansas, Louisiana, Oklahoma and Texas □ enough to last over100 years. Some really heavy hitters are pushing CNG, including the likes of Bill Clinton and T. Boone Pickins. Pickins has temporarily set aside his plan to build the world's largest wind farm in the Panhandle. He now plans to cut our foreign oil bill by one-third by filling America's highways with tens of millions of CNG-powered vehicles.

GREEN GASOLINE: A team led by George Huber of the University of Massachusetts – Amherst (UMass) have converted cellulosic biomass, like switchgrass, directly into gasoline components. The UMass researchers rapidly heated cellulose in the presence of solid catalysts that accelerated the reaction without being consumed in the process. Rapid cooling produced a liquid that contained many of the components found in gasoline. The entire process was completed in under two minutes, using only moderate amounts of heat. The compounds that were formed, like naphthalene and toluene, make up one-fourth of the chemicals in gasoline. The compound can be further processed to create green gasoline or used, as is, for a high-octane gasoline blend. The technology not only provides a method to quickly process large quantities of biomass, but theoretically, it does not require any external energy.

Huber says: “It is likely that the future consumer will not even know that they are putting biofuels in their car. Biofuels in the future will most likely be similar in chemical composition to gasoline and diesel fuel used today. The challenge for chemical engineers is to efficiently produce liquid fuels from biomass while fitting into the existing infrastructure today. We are currently working on understanding the chemistry of this process and designing new catalysts and reactors for this single step technique. This fundamental chemical understanding will allow us to design more efficient processes that will accelerate the commercialization of green gasoline."

John Regalbuto, who directs the Catalysis and Biocatalysis Program at NSF said: "Green gasoline is an attractive alternative to bioethanol since it can be used in existing engines and does not incur the 30 percent gas mileage penalty of ethanol-based flex fuel. In theory it requires much less energy to make than ethanol, giving it a smaller carbon footprint and making it cheaper to produce. Making it from cellulose sources such as switchgrass or poplar trees grown as energy crops, or forest or agricultural residues such as wood chips or corn stover, solves the lifecycle greenhouse gas problem that has recently surfaced with corn ethanol and soy biodiesel. Huber's new process for the direct conversion of cellulose to gasoline aromatics is at the leading edge of the new 'Green Gasoline' alternative energy paradigm that NSF, along with other federal agencies, is helping to promote."

Academic laboratories, small businesses and Fortune 500 petroleum refiners are pursuing green gasoline. They are designing ways to convert refineries to produce fuels, textiles, and plastics from either crude oil or biomass. DOD is a strong supporter of these efforts to produce jet fuel and diesel. NSF, DOE and the American Chemical Society recently released a report entitled: "Breaking the Chemical and Engineering Barriers to Lignocellulosic Biofuels: Next Generation Hydrocarbon Biorefineries." (4/1/08)

GEOTHERMAL POWER: Temperatures at the Earth's core reach 7000 degrees Centigrade; hotter than the surface of the sun. Geothermal energy could meet the world's annual energy requirements 250,000 times over, with zero impact on the climate or environment. MIT said that if 40% of the heat under the U.S. could be tapped, it would meet the nation's energy demands 56,000 times over. Which means if we could tap just one-hundredth of one percent, our energy needs would be well covered. Geothermal drilling is costly, but once operational, it is cheaper than biomass or solar power. In a place called Innamincka in South Australia's outback, Geodynamics is working to make geothermal power a major source for Australia's electricity grid. Tests have shown that the granite "hot rocks" 2.8 miles beneath Innamincka's red desert are the hottest, 250 degrees Centigrade, of their kind on earth. One cubic kilometer holds the stored energy equivalent of 40 million barrels of petroleum.

SOLAR: National Geographic says "the sun is a utopian fuel: limitless, ubiquitous, and clean." Solar energy potential is off the charts, The sunlight energy striking the Earth every 40 minutes is equivalent to the total global energy consumption for one year. Every hour the sun floods the earth with thermal energy equivalent to 21 billion tons of coal. A solar energy farm 100 miles by 100 miles in the Nevada desert (10% of the state) could meet the total energy requirements of the U.S. A 70 mile by 70 mile farm could do the same for Europe. There are no greenhouse gas emissions and no evironmental footprint beyond land use. According to the World Coal Institute, at the present rate of consumption, the world will run out of coal in 130 years, natural gas in 60 years, and oil in 42 years.

World energy consumption is 15 Terawatts (TW) (15X10^12 watts) per day. The total energy that strikes the Earth daily is 166 Petawatts (PW) (166X10^15 watts). Calculating that 50%

of this energy is reflected back into space or absorbed by clouds, the remaining 83 PW is more than 5000 times our present global energy consumption. Professor Derek of the University of Adelaide in Australia wants to use solar farms of mirrors to focus sunlight on collectors that boil liquid (water, oil, etc.) to create steam to generate electricity. Solar farms of one or two square kilometers could be built in the deserts of the Americas, Africa, Asia, Australia, and the Middle East. Abbott says we only have to tap 5% of solar energy at an efficiency of 1% to generate 5 times the world's current consumption. MIT's quantum-confined materials laboratory is experimenting with graphene, which is a single layer of carbon atoms. In a copper wire electrons travel at a snail's pace; in graphene they travel at nearly the speed of light. MIT believes graphene could provide the ultimate solar panel conductor; it is superconductive, cheap, and so thin it is transparent to light.

Starwarswise, vast arrays of solar panels in geosynchronous orbit could beam microwave energy back to antennas on Earth, where the waves could be converted to electricity. Geosynchronous orbit could keep the solar panels arrays in the sun nearly 24/7.

NANOTECHNOLOGY: The world is on the threshold of the greatest advance in the history of science and technology - the ability to build molecules atom by atom. These molecular building blocks can be arranged in any combination of patterns to produce every conceivable substance or device. This new technology - nanotechnology - will be an advance equivalent to a leap from the Stone Age to the Nuclear Age. How small is a nanometer? By definition, one nanometer is a billionth of a meter, but that's a hard concept for most of us to grasp. Here are some other ways to think about how small a nanometer is:

A sheet of paper is about 100,000 nanometers thick.

There are 25,400,000 nanometers per inch.

A nanometer is a millionth of a millimeter.

A micron is a millionth of a meter; a nanometer is a billionth. A typical human cell has a diameter of about 30 microns; a typical atom a diameter of about 0.3 nanometers. This is a linear difference of 100,000, and a volumetric difference of 1,000,000,000,000,000, the difference between a marble and a mountain.

The most advanced microchips, micromotors, microrobots, etc., will be a billion times the size of nanochips, nanomotors, and nanorobots. The difference is basic: in microtechnology, you are building-down, smaller and smaller; in nanotechnology, you are building-up, atom by atom. The advances that will be made in microtechnology pale by comparison with those expected with the arrival of nanotechnology. Molecular assemblers, built out of nanoscale components - nanoarms and grippers, driven by nanomotors and guided by nanocomputers – will build anything atom by atom, molecular by molecular. Nanorobot arms will be 50 million times shorter than human arms and thus will be able to move 50 million times faster. The nanocomputers will be smaller than a synapse and a million times faster. They will direct nanorobots to disassemble an object, record its structure, then assemble perfect copies. The end products will be remarkable:

> Aircraft, spacecraft, and cars built out of diamond-based structural material that is 50 times stronger, 14 times stiffer, and 90% lighter than aluminum.
>
> A nanocomputer equivalent to a modern mainframe computer would fit into one cubic micron; over 14,000 would fit into a human cell.

Nanotechnology will have an equivalent impact on medicine. The ill, the old, and the injured are all victims of misarranged atoms that can be rearranged by nanomachines. The solution to most medical problems involves selective destruction of bacteria, viruses, cancer cells, etc. Cell-repair machines the size of bacteria and viruses will conduct search-and-destroy operations against offending replicators.

Accelerators able to shoot femtosecond x-ray pulses (x-rays a quadrillionth of a second long) would be able to photograph chemical reactions as they occur. By making these pulses occur one after another, motion pictures of the reactions could be created. Such devices would permit us to examine in infinite detail the basic nature of all matter and provide a giant step on the road to nanotechnology.

Advances in chemistry and biotechnology will lead to enzyme-like and protein-like machines able to bind atoms and molecules together one at a time, building up larger and larger structures of any material in any shape. Motors, shafts, gears, bearings, etc., made of ceramics, diamonds, or plastics in tightly bonded sets of atoms will be assembled into machines with robot arms less than a tenth of a millionth of a meter long, building objects at a rate of millions of molecular operations per second. Building large objects one atom at a time would be slow, a fly contains about one million atoms for every second since the time of the dinosaurs. But nanomachines will be making copies of themselves at the same rate as bacterium, about one copy every fifteen minutes, just as "biology manufactures bacteria, butterflies, and buffaloes." With each replicator making more replicators, production rates increase exponentially; at the end of ten hours they would weigh more than a ton. An invention made on Monday could by the following Friday, be in mass production, with billions of copies fabricated.

Molecular machines will be dirt cheap, since they will be built out of elements like hydrogen, oxygen, nitrogen, carbon, silicon, and aluminum. Advanced automated design, engineering, and manufacturing will minimize labor costs. By proper arrangement of atoms, extremely strong materials can be made. For example, by arranging carbon atoms in straight chains, carbyne, which is ten times stronger than steel, can be made. Nanoassemblers could build engines out of carbon in its diamond form; they would be stronger than steel but lighter than wood. Such engines would be seamless and gemlike and have 90% less mass than a modern metal engine.

Nanotechnology could be the most powerful new force in history; the Industrial Revolution would pale by comparison.

"Self-replicating assemblers will enable us to make everything from microscopic computers to great star-ships for the price of crabgrass. The industrial revolution gave us a capital base that can be made to double in decades, with hard work; replicating assemblers will give us a capital base that can double in minutes, with no work at all. Today, economic and military affairs are stabilized by sheer sluggishness of production technology. To go from making a single computer to having millions - or from a single missile to mere thousands - takes years. A prototype embodying a technical breakthrough has no immediate revolutionary effect. But with production based on replicating assemblers, the time required to go from having a working prototype to having a million working units will shrink from years to days.

With nanotechnology we'll be able to make almost anything we want in any amount we want, and do it cheaply and cleanly. Poverty, homelessness, and starvation can be banished. Polution can be eliminated. We can finally open the space frontier, With the help of powerful AI systems, we'll be able to tackle more complex applications of nanotechnology, including molecular surgery to repair human tissue. And that can eliminate aging and disease. People everywhere struggle for greater wealth and better health. With these advances, we can have them - for all of us."

K. Eric Drexler
Engines of Creation

ROBOTICS: In the field of computers, Moore's Law correctly predicted CPU power would increase by a factor of 1000 every 15 to 20 years. Applying Moore's Law to robotics, every 15 to 20 years robotic technology will advance by a factor of 1000 or more. The human brain is thought to be able to process information at a rate of one quadrillion operations per second. Futurist Ray Kurzweil has predicted that by the 2020's it will be possible to construct a machine that is as intelligent as a human being. Lewey Gilstrap, author of the forthcoming book, *Machine Intelligence and Robotics,* believes we can begin building machines that can match or excel human intelligence within five years. He believes such machines could grow to be a $5 trillion to $10 trillion industry within ten years.
Robots will take over the workplace:

Nearly every manufacturing job will be robot.
Nearly every construction job will be robotic.
Nearly every transportation job will be robotic.
Nearly every home will have robot servants to do the housework and yardwork.
Robots will care for the elderly in their own homes.
Nearly every hotel and restaurant job will be robotic.
Most retail and wholesale jobs will be robotic.

Robots will be operating the factories, building the roads, constructing the buildings, and driving the trucks. Already, swarms of robotic aircraft are fighting the air war in Afghanistan. The next generation of bombers and fighters will be robotic; beginning with freighters, commercial aircraft will follow. Robotic waiters and retail clerks will bring a great improvement in service. Robotic K-12 teachers will significantly improve the quality of secondary education. Finally, we will be able to rid ourselves of the burden of the Department of Education and the all-powerful NEA. Agriculture will go totally robotic. Police and firefighters will be robots, perhaps ending charges of racism and police brutality; or perhaps not. Hospitals will cease being as dangerous as minefields and the ghetto, when operations, nursing, and sanitation are handled by robots.

GARBAGE: GM is teaming up with Coskata Inc. to mass produce ethanol from any kind of municipal waste. The process turns garbage, wood chips, garden waste, even old tires, into ethanol in two minutes flat. Argonne National Laboratory analyzed Coskata's operation and says the process yields 7.7 units of energy for every unit of energy put into the system.

The process uses less than one gallon of water to produce each gallon of ethanol, compared to 3-4 gallons of water per gallon of corn ethanol, and 44 gallons of water per gallon of gasoline. One ton of dry input yields 100 gallons of ethanol. Coskata plans to produce 100 million gallons of ethanol annually by 2010. The Department of Agriculture says the U.S. has the potential to produce one billion tons of biomass annually from garbage and agricultural waste. Using Coskata's process, this would yield 100 billion gallons of ethanol per year.

SUGAR TO GASOLINE: Nature is very good at turning sugar to ethanol, but not so good at turning sugar to hydrocarbons.

Amyris Biotechnologies of Emeryville, CA is using genetic engineering to turn sugar directly into hydrocarbon. Amyris adds genes to E.Coli and other microbes to get them to digest glucose and secrete gasoline. Amyris is being funded by the Gates Foundation, Khosla Ventures, Kleiner Perkins Caulfield & Byers. Amyris is building a plant to produce fuel in industrial quantities. It is talking to Brazilian sugar suppliers and to Costco and others about distribution.

JATROPHA: The African Jatropha plant favors hot, dry, tropical conditions. Plant it once and it will grow for 50 years on land that is like asphalt! Africa and India have huge areas of degraded and semiarid land that are ideal for its cultivation. Africa and India have the planet's cheapest and neediest labor. Jatropha yields per acre are comparable to switchgrass and miscanthus. Hundreds of millions of African and Indian farmers live on less than one dollar a day; jatropha could lift them out of poverty. In 2006, over 17,000 Indian farmers killed themselves to escape unpayable debts. One of DOE secretary Steven Chu's solutions to the energy problems is to create a "global glucose economy." Millions of acres of fast-growing crops like jatropha would be planted in the tropics. The plants would be converted to glucose and shipped around the world like petroleum is today. The glucose would be converted into biofuel and bioplastics.

CO_2 + SUNLIGHT = FUEL: Using sunlight, CO_2, and genetically engineered microorganisms, Cambridge, Mass-based Joule Biotechnologies plans to make liquid fuels and chemicals directly. The engineered microorganisms grow through photosynthesis in a brackish water solution, directly excreting fuel or commercial chemicals. No biomass feedstock, such as algae or other plants, are required, only CO_2, H2O, and the secret microorganism. Their Solar Converter is something like a solar panel with liquid inside. Going directly from CO_2 and sunlight to the end product is highly efficient. They claim the process can produce 20,000 gallons per acre per year. No agricultural quality land is required. A desert area the size of the Texas panhandle could produce enough fuel to replace all petroleum used in the U.S.

To get the large amounts of CO_2 required, Joule plans to capture the flue gas from large emitters like coal-fired power plants and cement factories. Algae bioreactors utilize the same approach of capturing industrial CO_2, but require the algae to be processed to extract the fuel, whereas, the Joule process produces the fuel directly. The company plans to set up a pilot plant to test its prototype Solar Converter in New Mexico in 2010, and plans to build a large industrial-scale facility in late 2011.

Dow Chemical plans to build a demonstration plant at Freeport Texas that can produce 100,000 gallons of ethanol per year. The layout consists of 3100 5X50 foot algae growing covered troughs. They plan to bypass the refining step and convert CO_2 directly into ethanol.

North Dakota alone has the potential to provide electric power for more than 25% of the country. The Dakotas have the potential to produce 50 million tons of hydrogen per year from wind machines; enough to power every car in America. Adding Texas, Kansas, and Montana would increase that output by 150%.

THE ASIAN CENTURY

The most important geopolitical economic story of the next 20 years will be the emergence of an Asian Confederation led by

China and India - Chindia. The emergence of these two colossi will shape the future of capitalism. China is the world's leading consumer of commodities – steel, coal, cement, grain, meat – and second only to the U.S. in oil consumption. It is building a gigantic new infrastructure of factories, power plants, dams, highways, railroads, skyscrapers, and military machine. China plans to quadruple the size of its economy in the next 20 years. China is building a 53,000 mile superhighway system; in three years it will have more miles than the U.S. national highway system's 46,000 miles. China is building a 16,000 mile high-speed (220 mph) national railroad network, that when completed in 2020, will be the largest, fastest, most technologically advanced rail system in the world. It is being built on 200,000 massive (800 ton) elevated buttresses lined up about 250 feet apart across the Chinese countryside, like a new Great Wall. India is connecting every village of over 1000 people to the national highway grid by 2010. China and India are rebuilding the World War II Burma Road as part of a New Silk Road that will run from the Persian Gulf through Bombay, Singapore, Hong Kong, and Shanghai.

Within five years, China will become the leading industrial power in the world. China's GDP is the fastest growing in the world. China is already a top world shipbuilder, electronics producer, and chemical manufacturer. It produces more than 50% of the world's steel and controls 90% of the world's rare-earth oxides. It's housing industry, auto industry, and stock market are booming. China has no capital gains tax; it's corporate tax rate is 15%. Until now, China has saved too much, consumed too little, and been too dependent on exports to fuel its economy. China is being converted from an export economy to a consumer economy, like the U.S. China's state mercantilism is more nationalist than socialist. Since 1999, China's known military expenditures have grown faster than its GDP - 15% to

19% annually. Military R&D spending growth is even greater. Pax Sinica's long-range goal: the yuan will replace the dollar as the world's reserve currency; Shanghai will replace New York and London as the center of finance; Mandarin will rival English as the Welt Sprache; Europe will become a quaint culture under glass like Athens and Rome.

Professor Guy Sorman in *Economics Does Not Lie,* outlines ten key economic principles, among these are:

Growth is the best measure of economic health

Creative destruction is the engine of economic growth

Per Dr. Sorman, China's rapidly growing GDP is the best measure of its economic health. This should be tempered somewhat by the fact that, thus far, this growth is largely confined to the 200 million Chinese clinging like crabs to the Eastern coastline. As to Sorman's second principle, America leads the world in creative destruction, driven by technical innovation and entrepreneurialism. America is unique in its enormous university/industry cooperative contrivance and its vast venture capital system.

Coal-fired power plants are the cheapest way to generate electricity. China has 25% of the world's coal supply. China gets 80% of its electricity from coal-fired power plants. China is building three new coal-fired power plants per week; so this percentage of electricity from coal is destined to grow exponentially. India gets 70% of its electricity from coal-fired power plants and also has a large building program. America has 25% of the world's coal supply; and gets 50% of its electricity from coal-fired power plants. America was scheduled to build 280 500-megawatt plants in the next 20 years; environmentalists have blocked this modest (compared to China) building program. It is not necessary to spend trillions of dollars developing "clean coal". America should copy the Chinese and build more coal-fired power plants to fuel more CO_2 fueled algae bioreactors; the more CO_2, the more biofuel. With thousands of coal-fired power plants fueling algae bioreactors, the U.S., China, and India would be awash in an ocean of algae biofuel! And in the process, oceans of CO_2 will be continually consumed!

China, India, and Korea have the world's fastest growing economies. As these giants achieve world economic domination, the rest of Asia will rush to jump aboard the juggernaut. Russia will be persuaded the confederation will be in the best position to exploit the enormous territorial and commodity resources of the vast Siberian Golconda. Siberia will serve the same purpose to this runaway Colossus as the American West did to us □ a giant treasure chest and pressure relief valve.

A new wave of agricultural colonialism is sweeping the globe. China and the Middle East petrobarons are buying up all the arable land on the planet, particularly in Africa. Led by China and Saudi Arabia, hordes of "carpetbaggers" are sweeping through Africa, Australia, Brazil, Ukraine, and the Philippines like they were grabbing the Reconstruction South. China is not just buying Africa, they are sending hordes of Chinese farmers to work the land. The one child per family rule does not apply to Chinese living abroad. As in the American West, the adventure of being pioneers in a new land will attract millions of young Chinese families, particularly from the less developed Western provinces. The long-range outcome is obvious; the intelligent, industrious, prolific Chinese will eventually own and run Africa.

PAPER EMPIRE

Money borrowed in the 1970s and 1980s was spent on factories, highways, defense, etc. Money borrowed in the 2000s was used by hedge fund managers and assorted "masters of the universe" Wall Street hotshots to maximize the leverage of their global speculations. Excess exceeded that of the Robber Barons. America's power brokers decided that finance, not manufacturing or high technology, would be the El Dorado of the 21st century. Paper entrepreneurialism became the ruling religion. Wealth could be maximized without the drudgery of creating or manufacturing anything.

By 2004 - 2006, financial services represented 20% of the GDP; manufacturing dropped to 12%. Hedge funds multiplied from a couple of hundred in the early 1990s to over 10,000 in 2007; accounting for 50% of the daily trading volume of the New York Stock Exchange. Utilizing exotic quantitative mathematics, they arbitraged the nooks and crannies of global finance. Enormous bank loans yielded twenty-to-one (and higher) leverage that transformed mere decimal point changes into financial bonanzas. In 2006, a Forbes survey found that the highest paid hedge fund managers made an average of $657.5 million per year; the median annual salary for an American bricklayer was $44,833. Ergo, the hedge fund manager made as much as over 14,000 bricklayers, for producing absolutely no useful end product. In the economic expansion of 2001 - 2007, poverty rose and median income for working age households fell, the opposite of what normally occurs during an expansion. In America, executive compensation has become obscene, CEO/journeyman pay ratios of 500 to 1 are common; in the rest of the world, 25 to 1 is the norm.

Mortgage companies and bank mortgage departments offered new and exotic interest-only mortgages, payment-option adjustable-rate agreements, piggyback loans, etc., etc., all buried in pages of tiny indecipherable print (fine print is never good news). Buyers were steered to the most complex and toxic instruments, which yielded the highest fees. The economy came to be driven by a financial sector dominated by black box and algorithm vendors. By 2010, algorithmic trading was projected to account for half of all trading in the U.S. equity market.

Financial Times associate editor Wolfgang Munchen wrote: “The reason why this crisis is so nasty has to do with the deep inter-linkages within the credit market and the real economy. Take for example, a synthetic collateralized debt obligation, one of the most complicated financial instruments ever invented. It consists of a couple of credit default swaps, credit linked notes, total return swaps, all jointly connected in a wiring diagram that looks as though the structure was about to explode.” *Bad Money* by Kevin Phillips, Viking, NYC, 2008 (p.101).

Wall Street operates in an environment of minute by minute cut-throat competition to produce. It is a world completely divorced from American society – it might as well be on Mars. The general welfare of America never enters the equation. Despite the current economic calamity, Wall Street has resumed business-as-usual with a vengeance. In 2006, Goldman Sachs allocated $16.7 billion for year-end bonuses; Goldman Sachs’ bonus pool for 2009 is $20 billion. Currently, a $600 trillion market in unregulated derivatives is lurking offstage. Tens of trillions of dollars of new toxic instruments are poised ready to be added to the tens of trillions already polluting the planet. Unchecked, total economic collapse is inevitable.

SCIENTISTS AND ENGINEERS

K - 12 education in America is abysmal. Math and science are taught by people not educated in the subjects. Majoring in science and engineering (S&E) is tough; K - 12 is managed and taught by education majors. George Mason University professor Walter Williams says: "Students who have chosen education as their major have the lowest SAT scores of any major. Students who have an education degree earn lower scores than any other major on graduate school admission tests such as GRE, MCAT or LSAT. Schools of education, either graduate or undergraduate, represent the academic slums of most any university. They are home to the least able students and professors. Schools of education should be shut down." In the 2000's, our universities "best and brightest" went for where the big bucks were – the law and Wall Street. In their view, only the rubes and round haircuts went for engineering. The "best and brightest" became "masters of the universe" and wrecked the world economy. With Wall Street now held in the same exalted esteem they enjoyed in the 1930s, perhaps some of these worthies could be enticed to pursue careers in industries where useful products are made. Or perhaps not. Wall Street appears to be back in business at the same old stand.

Armies of scientists and engineers are a vital component of the new technologies we must develop. China is graduating 500,000 engineers per year; India 350,000. America must maintain its lead in R&D. If researcher salaries were higher we would not have to depend on foreigners to fill our graduate programs. If we paid S&Es as well as we pay lawyers and investment bankers, students would be tripping over each other to study the sciences. To truly stimulate the economy, America must go back to being a manufacturing country. Economics 101 teaches that economic growth is driven by innovation and technological advances, both rooted in manufacturing. Manufacturing accounts for two-thirds of private sector R&D. Currently, manufacturing accounts for 12% of our GDP and employs 12 million workers; down from twice that in the 1970s.

America must end "invented here, industrialized elsewhere." A galaxy of marvelous new technologies developed with American taxpayer dollars are ready to burst upon the world market. America must resurrect its manufacturing base or the new factories and high-paying jobs from these emerging technologies will end up in foreign lands. America must find a way to bridge the disconnect between R&D and commercialization.

RESURRECT HENRY FORD

Henry Ford was a long-term thinker and America is in competition with long-term thinkers in Asia. Every year Henry Ford made the Model T a little bit better and a little bit cheaper, and every year he sold more Model Ts than the year before. When the demand for his Model Ts went up each year, Henry did not raise the price in order to raise the pay and bonuses of Fomoco's top brass. Ford did not send corporate raiders into the marketplace searching for companies to plunder for their assets (buy, strip, and flip). Instead, Henry cut the price of the Model T, so that more American working people could own one. Henry put the world on wheels and made Ford the biggest company in the world. This was an age when companies like Ford and Duesenberg were run by car lovers competing to grow the company and improve the breed.

Greed, instant gratification and bottom-line, short-term thinking have wrecked Wall Street and Detroit. Industry today is too often run by lawyers, accountants, and assorted raiders and plunderers who have no interest in growing and nurturing a business to build a better product and provide long-term employment. We are in competition with long-range planners. America's corporate and political heavy-hitters think in terms of the next quarters earnings or the next election. In Asia, our competition thinks in terms of the next century.

PROGRESSIVE INSURANCE AUTOMOTIVE X PRIZE (PIAXP)

In 1927, Charles Lindbergh flew solo nonstop across the Atlantic to win the $25,000 Raymond Orteig Prize; within a year the number of pilots in the U.S. tripled; the number of aircraft quadrupled. The PIAXP is a set of competitions, programs, and events that X-Prize Foundation chairman Peter Diamandis hopes will "inspire a new generation of super-efficient vehicles that help break America's addiction to oil and stem the effects of climate change." 95 teams have registered for the competition. The objective of the competition is to design, build and race super-efficient vehicles that will achieve 100 mpg(e) and can be produced for the mass market. PIAXP selected 100 mpg as the happy medium; the perfect being the enemy of the good. For example, at 20 mpg, it takes five gallons to go 100 miles. At 100 mpg, it takes one gallon, so you save four gallons. At 200 mpg, it takes 1/2 gallon, so you save only 1/2 gallon more. PIAXP says at 80 mpg, a vehicle achieves 94% of the energy savings of a 100 mpg vehicle.

There is a Competition Division and a Demonstration Division. In the Competition Division there are two vehicle classes: Mainstream and Alternative. The Mainstream class has a prize of $5 million. The Alternative class has two separate prizes of $2.5 million, one for side-by-side seating and one for tandem seating. PIAXP also has a Demonstration Division for well-established automotive companies already selling vehicles in the U.S. or the E.U. at rates of over 10,000 cars per year. The Demonstration Division was set up to allow car companies to demonstrate existing or planned high-efficiency vehicles like the Chevy Volt. These vehicles will participate in the same events as the Competition Division, but will receive no prizes.

The mainstream class vehicles must seat four or more adults, have four or more wheels, have 10 cubic feet of cargo space, maintain 65mph on a 4% uphill grade, accelerate from 40mph-60mph in under 9 seconds, and have a minimum range of 200 miles without refueling or recharging.

The alternative class vehicles must seat at least two adults, no minimum number of wheels, accelerate from 0 - 60 mph in 18 seconds or less, and have a minimum range of 100 miles without refueling or recharging.

Both classes of vehicles must be equipped with standard automotive features, including enclosed cabins with windshield and windows, wipers, washers, headlights, brake lights, horn, rear and side-view mirrors, and seat belts.

The competition events are scheduled to run from Spring 2010 to Fall 2010. The goals are to test vehicles under myriad and real world conditions over an extended period to demonstrate durability. The purpose is maximum local and international media exposure to educate the public about fuel economy, emissions, and alternative fuels. The original X-Prize received $500 million worth of media coverage worldwide and over $5 billion in web media impressions. PIAXP expects an increase in coverage of 400% for this event, or $2 billion in media coverage and $20 billion in web media impressions. Coverage will be extensive: Magazines, TV Shows, Documentaries, Premiers, Racing Events, International Auto Shows, Le Mans, Monaco Grand Prix, Concours d'Elegance, etc. The contest is open to all comers and should inspire a galaxy of exciting new concepts in the tradition of Ultralite, Hypercar, and the Autonomy. Examples are entries like Aptera, Loremo, and Velozzi.

Munich manufacturer, Loremo AG, (Low Resistance Mobile) introduced its new model 'GT' in 2009. It is a four-passenger vehicle, weighing in at 1200 pounds, powered by a two cylinder, 46kW/60hp gasoline engine. It is said to be capable of accelerating from zero to 100 km/hr in 9 seconds and get 78 mpg. Loremo AG lists the 'GT' at 24,000 Euros. They have another model 'LS' that gets 120 mpg, but takes 20 seconds to go from 0-100 km/hr. Since gasoline costs $7 a gallon in Germany, they are willing to trade lower acceleration for higher mileage. The Loremo EV is an electric version; it weighs 1300 pounds, has a 20kW Li-Ion battery, goes 0-100 km/hr in 15 seconds, and has a range of 93 miles. All three models have a Cd of 0.214. There will be a special edition Loremo for the AXP.

Three San Diego engineers led by Steve Fambro set up Aptera Motors to build the Aptera. It is a two-seat, three-wheeled machine that looks like a dolphin and has the drag coefficient of a gnat. The shape was optimized by extensive virtual wind tunnel testing. The body is constructed of composites using Aptera's patented Panelized Automated Composite Construction (PAC2) fully automated manufacturing process which greatly reduces time and cost. Because it has three wheels, most states classify the Aptera as a motorcycle, which means safety and emission tests are not mandatory. Notwithstanding this, the Aptera 2e and 2h models incorporate a Formula One type "crash box" safety cell for passenger protection.

The Mk-1 prototype model was powered by a 12 hp diesel engine and a 25 hp permanent magnet DC motor, backed up by a supercapacitor and regenerative braking. It weighed 850 lbs, had a Cd of 0.11, low rolling resistance tires, had a top speed of 95 mph (electronically limited), and got 330 mpg at 65 mph. Aptera utilizes a patent-pending hybrid technology that permits off-the-shelf engines and electric motors to be seamlessly integrated at a very low cost. The Aptera 2e, is all electric, weighs 1500 lbs, has a Cd of 0.15, is powered by a lithium iron phosphate battery, and has an all electric range of 120 miles. The Aptera 2h is a series hybrid PHEV. It has a small gasoline engine to recharge to recharge the battery and supercapacitor. For a 120 mile trip after a full charge, it gets 330 mpg; its five gallon fuel tank gives it a range of 600 to 700 miles. Aptera says that 99% of Americans drive fewer than 120 miles per day.

Led by CEO Roberto Jerez, a blue-ribbon team of scientists, racing engineers and builders have designed the Velozzi, a fabulous high-performance PHEV that will go from 0-60mph in 3 seconds, have a top speed of 200mph, and get 100mpg. Velozzi has teamed with Saminco, Inc. of Ft. Myer, FL, manufacturer of premier automotive electric drive systems. A Saminco car holds the world's land speed record for an electric vehicle – 321mph. The Velozzi is driven by an AC electric motor, powered by a lithium-ion battery pack and supercapacitors, recharged by a multi-fueled micro-turbine. It will be constructed of space-age lightweight carbon fiber nanotube composites. The price goal is under $30,000. Team Velozzi is planning to enter two cars in the AXP competition. The two-seat Velozzi in the Alternative class; a four-seat Solo in the Mainstream class.

The Solo is a crossover designed to have the performance of a high-end exotic. The Velozzi and the Solo will be the first production PHEVs to use multi-fuel micro-turbine battery charger to recharge a pack of supercapacitors and Li-Ion batteries. The Velozzi vehicles will have the unique ability to reverse polarity and become power generators able to serve as home auxiliary power units (APU). Velozzis will have regenerative braking and driver transparent embedded computers to control all functions. Velozzis will be the first production vehicles to use carbon fiber nanotubes in their construction. Carbon nanotubes have the greatest tensile strength of any material known, able to resist 100 times more strain than typical structural steel.

Seventeen global tier-one OEM suppliers have joined Velozzi's world-class team. Velozzi provides expertise in hybrid systems and automotive design and construction, Saminco are experts in traction controls, Weissmann R&D in transmission know-how, and Bayer Material Science (BMS) brings broad spectrum expertise in materials. BMS offers polycarbonate automotive glazing, polyurethane composite parts, and carbon fiber nanotubes for major body components. BMS also provides raw materials for waterborne polyurethane coatings and adhesives, bio-based foam for seating and headrests.

The Velozzi Supercar is a high-performance limited production sports car targeted at the affluent auto enthusiast. The Solo is a four passenger crossover vehicle aimed at the mass market. All Velozzi vehicles are PHEVs powered by lightweight lithium-ion batteries, backed-up by lightweight supercapacitors. The battery packs and supercapacitors are recharged by multi-fuel micro-turbines and regenerative braking. The multi-fuel micro-turbines can burn any grade gasoline or diesel fuel, ethanol, methanol, bio-diesel, CNG, hydrogen, etc., etc. The Supercar will be powered by a 770HP AC induction electric motor, will accelerate from 0-60mph in 3 seconds, and have a top speed of over 200mph. The Solo will exceed 100mpg at a sustained cruising speed of 70mph. Its 250HP AC induction electric motor will accelerate from 0-60mph in under 6 seconds and reach a top speed of 120mph. Fully charged the battery pack range is over 200 miles; with a fully fueled range of nearly 1000 miles.

The micro-turbine charger maintains the optimal power level to extend battery life. The supercapacitors are employed for their rapid charge and discharge cycles which insulate the more delicate batteries. Bayer's cutting-edge lightweight polymers provide a level of safety usually seen only in racing vehicles. The combination battery pack and supercapacitor provides the AC induction motors with a tremendous amount of power; this power needs to be managed by a properly designed transmission. The transmission is typically the weakest component of most electric vehicles. Velozzis are equipped with transmissions designed by Weissman R&D. Over a 46 year period, Weissman R&D transmissions have won ten FIA F1 championships, five 24 hour LeMans, four Indy 500s, and set over 12 world speed records. Compact and lightweight the Weissman R&D unit designed for Velozzi has umparalleled performance and longevity. To extend the operational lifespan of the battery pack and other key components, Velozzi has a proprietary software management system (Velozzi SMS) run by a topflight team of software experts.

Velozzi is scheduled to go into mass production in late 2011 to early 2012.

Henry Ford changed the world with the Model T, Volkswagen with the Bug, and Tata Motors is about to with the Nano and the Indica EV. The world is on the threshold of two game-changing developments - EEStor supercapacitors and Tata low cost cars. The Tata Nano was designed to be the least expensive car in the world, with a starting price of Rs100,000 (1 lakh), approximately US$2000. When introduced in July 2009, the starting price was Rs115,000, US$2421. Nano means one-billionth - colloquially tiny.

The Tata Nano is a rear-engined, 4-door, 4-seat, 1300lb car powered by an aluminum 2-cylinder, 35hp, 624cc SOHC Bosch gasoline engine that gets 66mpg. The Nano Europa has a larger body, larger 3-cylinder engine, and an ABS system. A 690cc diesel version is reported to be under development. The Indica EV is designed to be the world's cheapest electric car. It is being developed by Miljobil Grenland AS in Norway. It will have an electric range of 125 miles, a top speed of 93mph, and is scheduled to go into production by the end of 2009. The British government gave Tata Motors a big loan to build it in a West Midlands factory.

In 2008, Warren Buffett's Berkshire Hathaway tried to buy 25% of BYD Company. CEO Wang Chuan-Fu declined that offer, but sold Buffett 9.89% for $230 million. BYD is located in Shenzhen, just across the river from Hong Kong; Shenzhen is the fastest growing city in the world. BYD employs 130,000 people in 11 factories, eight in China, and one each in India, Hungary, and Romania. They build cars, batteries, and solar energy systems. Berkshire Hathaway thinks BYD has the potential to become the world's largest builder of electric cars, batteries, and solar energy systems. BYD is one of the top four manufacturers in the world of rechargeable batteries: Li-Ion, NiCad, and NiMH. They are the largest manufacturers of these batteries in China. BYD wants to make its batteries 100% recyclable and have developed a nontoxic electrolyte fluid. BYD is working on a low-cost lithium ion ferrous phosphate battery. BYD's regiments of scientists and engineers (10,000+) are also working on rooftop photovoltaic panels with built-in batteries to store the energy when the sun is not shining.

Like Henry Ford's River Rouge, BYD is a vertically integrated operation; it manufactures all its components: engines, electric motors, bodies, electronics, lights, airbags, seatbelts, etc., etc. BYD is the first large auto builder to put a PHEV on the market – the F3DM (dual mode), a year ahead of GM's Volt or Toyota's plug-in Prius. The F3DM will have an electric only range of 62 miles and sell for $22,000. Also scheduled for 2009 is the e6, a 4-door, 5-passenger crossover car with a range of 250 miles, 0-60 in 8 seconds, top speed 100mph, 10 minute recharge to 50% power.

The BYD vs Tata race should be *very interesting!*

SKUNK WORKS

We are witnessing a fundamental change in the automobile industry. As Andy Grove said this is what happened in the computer industry in the 1980s and 1990s. Previously, the pioneering giants like IBM, produced their own mainframe computers that used proprietary hardware and software. IBM's sales force marketed these complex and expensive products. IBM ran everything – R&D, manufacturing, marketing. IBM was a "vertical" structure. Ford Motor Company was the same. Ford mined its own iron ore, produced its own steel, built its own engines and components, fabricated its own cars. At one point, it even grew its own rubber in Brazil.

The PC brought about a fundamental change in the computer industry. The industry began to use common hardware components (microprocessors) and packaged software; marketing was handed over to independents. The industry was transformed into a "horizontal" structure. Some companies build the components, others integrate them, others market them. Electric cars will transform the automobile industry from "vertical" to "horizontal". The electric car relies on a few fundamental components - batteries and supercapacitors, electric motors, and electronic control systems. There will be a worldwide competition to improve and manufacture these basic components. Anyone will be able to buy and assemble these components into a finished car - a kit car.

In the early 1900s, there were over 2000 car manufacturers in the U.S. The collapse of the Big Three combined with the emergence of new technologies like lithium-ion batteries and supercapacitors, carbon-fiber materials, biofuels, smartgrids, etc. will lead to the rebirth of the American auto industry and an enormous economic boom. Low volume construction technology, as espoused by the Rocky Mountain Institute, will permit thousands of small "skunk works" manufacturers to emerge. Examples of the potential of "skunk works" operations abound.

GM's Ultralite was designed and built by GM's Tech Center and Burt Rutan's Scaled Composites in 100 days. The whole job was done by 50 people at a cost of $4 million to $6 million.

Boeing's X-45C UCAS was built by fewer than 20 people with no hard tooling. It has an aluminum substructure with an all-composite skin in three parts the fuselage and two outer wings. All clips and brackets are bonded into the structure.

Andy Grove's program to convert 10 million (out of 80 million) PSVs (Pickups, SUVs, Vans) into PHEVS in four years, will lead to hundreds of shops becoming car builders.

The Automotive X Prize contest will jump-start hundreds of "skunk works" all over America.

The kit car industry, already healthy, will go exponential.

SET AMERICA FREE

Set America Free estimated we could have total energy independence in four years for $12 billion; which was about the cost of conducting the Iraq War for 40 days. (The Congressional Budget Office estimates the cost at $9 billion per month.) The average American driver will rarely need to buy a gallon of gasoline. Vast new high-tech industries will be operating. Millions of new high-paying jobs will have been created. All this is easily, quickly, and cheaply available if America sets up a Manhattan-type project to pursue the program advocated by Set America Free (www.setamericafree.org.) Set America Free (SAF) is promoting a comprehensive program of energy independence. The movement has attracted what ex-CIA director

James Woolsey calls: “a coalition of tree-huggers, do-gooders, sodbusters, evangelicals, cheap hawks, and venture capitalists – and Willie Nelson.” SAF’s estimate seems low. But if the cost of energy independence is several hundred billion dollars, it’s chump change to achieve the goal. We are currently spending $10 billion a month on imported oil.

Three of the many technologies SAF are pushing are plug-in hybrid vehicles (PHEV), flexible-fuel vehicles (FFV.) and alternative fuels. They say we can get all the fuel we need from coal, oil shale, tar sands, and a variety of agricultural products such as miscanthus and switchgrass. Estimates vary widely, State Farm Insurance estimates the average U. S. driver drives between 12,000 and 15,000 miles per year, which works out to 33 to 40 miles per day. DOE says 50 percent of U. S. drivers drive fewer than 26 miles per day and 80 percent drive fewer than 60 miles per day. Set America Free says 50% of U.S. cars are driven 20 miles per day or less, and that a PHEV with just a 20-mile range battery would reduce fuel consumption by about 85%. Alliance Bernstein researchers estimated 40% of Americans travel 20 miles or less per day; 60% travel 30 miles or less per day. Driving a conventional car costs 15 to 20 cents per mile; PHEV all-electric travel costs 2 to 4 cents per mile. The commuting “sweet spot” appears to be about 25 miles. A PHEV driven a mix of 12,000 city/highway miles a year, consumes 1840kWh to 2477kWh of electricity, depending on battery size. This is equivalent to the energy used in 3 to 5 months by an electric water heater in a three-person household (based on 540kWh/month). Owners of PHEVs with a 20 to 30 mile electric range would rarely have to buy any fuel. Electric vehicles represent the largest shift in automotive technology since the introduction of the internal combustion engine (ICE).

In 1996, GM introduced the EV1. It had a range of 55 to 95 miles before the batteries needed recharging. Its lead-acid (PbA) batteries took 6 to 8 hours to recharge. Thus, the EV1 could have met the needs of 80 percent of U. S. drivers. Burt Rutan happily drove a EV1 for years, before GM recalled and destroyed nearly all of them. GM CEO Rick Wagoner said his worst decisions were: "Axing the EV1 electric car program and not putting enough resources into hybrids." The EV1 used lead-acid batteries, the Toyota Prius uses more powerful nickel-metal hydride (NiMH) batteries; lithium-ion batteries (Li-ion) are much more powerful and quicker to recharge than either of these. Lithium-ion batteries recharge in four hours versus six to eight hours for lead-acid batteries. With high volume production, the cost of Li-ion materials are expected to decrease to less than the cost of NiMH materials. Prof. Andy Frank of the University of California at Davis has been building PHEVs for a decade, his research of PHEV technology indicates that the average person driving 40 miles per day or less would not use any gasoline at all, except on weekend trips and vacations across country. Dr. Frank said that at six cents per kilowatt hour, the electricity replacing gasoline costs about two cents per mile versus fifteen cents per mile using gasoline. In 2005, the U.S. electrical grid produced 4055 billion kwh. DOE's Pacific Northwest National Laboratory estimates the U.S. could meet the electrical needs of 73% of its 250 million cars and trucks with the existing grid. DOE compared greenhouse gas emissions of coal-fired power plants with those of ICE engines, and found that switching to PHEVs would yield a net reduction in greenhouse gas emissions of 27% per car. The Environmental and Energy Study Institute estimated that replacing conventional cars with PHEVs with a 20-mile electric range would reduce carbon emissions by up to 60% per vehicle.

The U. S. has the world's largest proven reserves of coal, convertible to trillions of barrels of methane or diesel oil. China may have an even larger supply of coal. China is building three new coal-fired power plants every week (not one, as is usually reported). China is pushing coal-to-liquids technology. America has 16,000 square miles of oil shale in Colorado, Utah, and Wyoming, convertible to trillions of barrels of oil; they contain triple the proven reserves of Saudi Arabia and could meet 25% of the current U.S. demand for 400 years. Canada has tar sands convertible to trillions of barrels of oil. The tar sands in Canada can produce an amount of oil equivalent to all of Saudi Arabia's proven reserves. Tar sands around the world can meet the planet's petroleum needs for hundreds of years. The U. S. has about 275 billion tons of recoverable coal reserves, about 25% of the world's total. The energy content of the U. S. coal reserves is four times greater than the recoverable reserves of Saudi Arabia; it exceeds that of all the world's known recoverable oil reserves. With current technology methanol is being cleanly produced from coal for less than 50 cents a gallon.

The road to the current energy mess goes back a long way. To give just two examples, between 1922 and 1955, GM, Standard Oil, and Firestone tire company bought up the nation's principal electric streetcar companies. By 1946, using a front company named National City Lines, they had bought and closed down the streetcar companies in 80 major cities, replacing the popular streetcars with fleets of pollution-belching buses. Back in 1964, President Lyndon Johnson slapped a 25% tax on all imported light trucks to retaliate against Europe for restricting the import of American frozen chickens. The U.S. auto industry seized the opportunity to launch a massive campaign to promote their trucks and utility vehicles. For decades they fought off (or bought off) government regulators attempts to push fuel efficiency by claiming it would hurt farmers and construction workers. As a result, Congress exempted light trucks from the 1975 Energy Policy and Conservation Act that set fuel economy standards. The Clean Air Act Amendment of 1977 permitted light trucks to emit two to five times as much pollutants as autos. In 1978, Congress exempted light trucks from the tax on gas-guzzlers and luxury cars. These events launched the sweetest money making machine in the history of the industry.

It was a glorious rebirth of the old Chevrolet/Cadillac story as related in John DeLorean's book, *On A Clear Day You Can See General Motors*. At that time, it cost GM only $300-$400 more to build a Cadillac DeVille than it cost to build a Chevrolet Caprice; yet the DeVille sold for $3800 more than the Caprice. The SUV and pickup truck market is the same old story. Detroit can sell a pickup truck for $20,000 and make handsome profit. As was the case with the Chevrolet/Cadillac, the industry can build an SUV body for little more than it costs to build a pickup truck body. The SUV body is mounted on the pickup truck chassis and the resulting SUV can be sold for $40,000 to $50,000. Pretty neat, eh?

Henry Ford put America and the world on wheels by improving his car every year and lowering its price every year. He was on a crusade to make his cars cheap enough that every workingman could afford one, and good enough that every workingman would want one. After Ford, the industry fell into the hands of the "bean-counters" and "bottom-liners." In the 1920s, GM's Alfred Sloan developed an ingenious plan to produce a spectrum of cars ranging in price from the Chevrolet to the Cadillac. As we've seen, it cost GM little more to build the Cadillac than the Chevrolet. The brilliance of the plan was to sell the American workingman on the idea of spending his lifetime climbing the social ladder – Chevrolet, Pontiac, Oldsmobile, Buick to Cadillac, to show the world he had finally "arrived." "Look Ma, top of the world!" The housing industry took the same route – "Starter-House" to "McMansion."

AMERICAN CENTURIONS

The world does not have an energy crisis; it has a petroleum crisis. And the crisis is not due to a shortage of petroleum, at least, not yet. The crisis is due to the fact that most of the world's petroleum supplies are in the hands of the world's latest crop of crazies. In the 20th century, we had lunatics like Hitler, Stalin, and Mao trying to force myriad brands of totalitarian insanity on the world. In the 21st century, we have a new crop of lunatics trying to ram totalitarian insanity down the throat of the world.

As outlined in Schultz and Woolsey's paper, *Oil and Security*, the world is facing a greater threat from radical Islamists than it did from the Soviets during the Cold War. Al Qaeda calls oil the "umbilical cord and lifeline of the crusader community." Two-thirds of Saudi oil goes through one processing plant and two terminals. Al Qaeda continues to try to destroy the sulfur-cleaning towers at Abcaig, the world's largest oil production facility, in northeastern Saudi Arabia. If they succeed, it would take six million barrels of oil a day off the world market for a year or more. This would drive the world price of oil to bin Laden's goal of $200 a barrel and trigger a worldwide economic depression. U.S. refineries are concentrated in a few vulnerable places along the Gulf Coast. The 800-mile Trans-Alaska Pipeline is stark naked to attack.

America consumes approximately 21 million barrels of oil a day; two-thirds (14 million barrels) of which are imported. America spends $10 billion a month; $400,000 a minute on imported oil. Five countries exported over 1.00 million barrels of crude oil per day to the U.S. These five countries accounted for 72% of U.S. crude oil imports in February 2007. These countries were Canada (1.838 million barrels per day), Mexico (1.358 million barrels per day), Saudi Arabia (1.185 million barrels per day), Venezuela (1.115 million barrels per day), and Nigeria (1.061 million barrels per day). Approximately 85% of U.S. Mid-East oil imports come from Saudi Arabia. Saudi Arabia's GDP soars every time the price of oil goes up. America pays two to three times as much to maintain military forces in the Gulf as we pay to buy oil from the Gulf.

Saudi Arabia has spent over $100 billion in the past 30-odd years funding Wahhabism throughout the world. Wahhabism is fanatically hostile to Shia and Suffi Muslims, Jews, Christians, Western civilization, women, and the modern world. Should these people succeed, it is their intention to take the whole world of Islam (and everybody else) back to the seventh century. Woolsey says the cash register at your local gas station is the collection box for Al Qaeda. "We're paying for both sides in this war, and that's not a good long-term strategy," he says. When some of America's top foreign policy experts were asked by *Foreign Policy* magazine what should be America's top priority in fighting terrorism, 82% said ending dependence on foreign oil.

Based on the figures cited above, we are importing 2.300 million barrels a day from Saudi Arabia and Venezuela combined. As noted earlier, Set America Free has said that 50% of U.S. cars are driven 20 miles per day or fewer, and that a PHEV with just a 20-mile range battery would reduce fuel consumption by approximately 85%. A 20-mile electric only range can be achieved with current state-of-the-art batteries. Thus, we can very quickly free ourselves from funding known enemies.

Aside from vast new discoveries of petroleum and natural gas, there's enough coal, oil shale, and tar sands, all convertible to oil, to keep the world supplied with fuel for hundreds of years. Then there's cellulosic ethanol. All over the planet, hundreds of millions of acres of unused land could be planted in crops like miscanthus and switchgrass, that could supply the world with renewable energy in perpetuity. The African Jatropha plant may be the next big biofuel. The plant favors hot, dry, tropical conditions. Africa and India have huge areas of degraded and semiarid land that are ideal for its cultivation, and have the planet's cheapest and neediest labor. Biodiesel entrepreneurs are buying up land and starting plantations.

Oil is a fundamental component of our economy and nothing but trouble. But it may be a blessing in disguise. It may be the catalyst we need to force us to revitalize our rural economy, provide energy security, lower fuel prices, open the door to new technologies, jump start legions of new entrepreneurs, slash the jugular of Middle East terrorist funding, and in the process, solve our unemployment problem. It took America only 12 years to switch from horses to cars and from steam to diesel-electric locomotives. Fifteen years from vacuum-tubes to transistors and from black and white TV to color. Auto bodies went from 85% open and made of wood in 1920 to 70% closed and made of steel in 1926. We can do it.

Farming was once considered the only honorable and respectable occupation for a Roman gentleman. Rome was a nation of yeoman farmers, initially, only property owners, like farmers, could serve in the Roman legions. If a soldier fought well he could become a Centurion. Cato said that when the forefathers "wanted to say that a man was good, their highest compliments were to call him 'a good farmer and a good husbandman'." "It's from the farming class that the bravest men and sturdiest soldiers came - - - their offspring are the strongest men and bravest soldiers; their profit the truest, safest, least envied; their cast of mind is the least dishonest of any." Marcus Porcius Cato of Rome, De Agricultura (circa 160 B.C.)

Jefferson, in Notes on the State of Virginia (1785) wrote: "Those who labor in the earth are the chosen people of God." In a letter to John Jay (1785), Jefferson wrote: "Cultivators of the earth are the most valuable citizens. They are the most vigorous, the most independent, the most virtuous, and they are tied to their country and wedded to its liberty and interests, by the most lasting bonds. As long, therefore, as they can find employment in this line, I would not convert them into mariners, artisans, or anything else." Jefferson dreamed of a nation of yeoman farmers.

In 1907, President Theodore Roosevelt said: "Nothing is more important to this country than the perpetuation of our system of medium-sized farms worked by their owners. We do not want to see our farmers sink to the condition of the peasants of the old world, barely able to live on their small holdings, nor do we want to see their places taken by wealthy men owning enormous estates which they work purely by tenants and hired servants."

The area of the U.S. is approximately 3,619,000 square miles; there are 640 acres per square mile; thus, the U.S. has about

2,316,160,000 acres. About 400 million acres are in cropland production (corn, wheat, soybeans, sorghum, alfalfa, hay, etc.) About 255 million acres are dedicated to animal feed production. About 350 million acres are rangeland. About 280 million acres are pastureland suitable for mechanical harvesting.

The U.S. consumes about 140 billion gallons of gasoline each year. Fifty million acres of land planted in prairie grass like miscanthus or switchgrass could produce enough ethanol to meet our requirements. Fifty million acres is about the size of South Dakota. That's about the amount of acreage in the soil bank (the Conservation Reserve Program (CRP) that the government pays farmers not to cultivate in order to prop up farm prices. At present, farming accounts for 10% of the GDP, yet employs fewer than 1% of the national workforce. Between 1948 and 2004, total farm production increased 166%, but productivity improved so much that only 25% as many farmers were needed. In 2005, 82% of U.S. harvested acreage was in four crops – corn, soybeans, hay and wheat. The U.N.'s Food and Agriculture Organization (FAO) is pushing conservation of topsoil, greater crop diversity, and crops that require little or no tillage. Switchgrass is tailor-made to meet all these requirements.

Once a mighty ocean of wild grass marched across the millions of acres of our Great Plains. It stood ten feet tall and gripped the earth with roots ten feet deep. It was switchgrass (Panicum virgatum), a fast growing grass found in North and South America and Africa. Switchgrass is a tough, hardy perennial; ready for harvest two years after planting; it can be harvested annually or semi-annually; it can be harvested for ten years or more before replanting is needed. Switchgrass is a perennial that grows in a wide variety of soils, needs no tillage, little fertilizer, water, or pesticides; it will grow in more parts of the country than corn.

It has a beautiful little purple and orange flower, like the Bird of Paradise. It reduces nitrogen runoff from fertilizers, increases carbon content in soil, reduces topsoil erosion, and provides excellent shelter for birds. Some Great Plains prairie grasses have been hayed for 75 years with no fertilizer and no decline in yield. The deep root systems of switchgrass and miscanthus deliver many important environmental and agricultural benefits. Switchgrass, covering the 21st century version of the Great Plains could produce a variety of fuels and chemical by-products. As it once fed tens of millions of buffalo; it could soon fuel tens of millions of cars, spin power turbines, and supply chemicals for American industry.

We are being offered a golden opportunity to launch a new era of eco-capitalism in rural America. Our rural economy can play a key role in putting America on the road to secure and sustainable energy independence, by producing and processing biofuels and bioproducts. Dedicated energy crops such as switchgrass, miscanthus, jatropha, and fast growing trees will increase farmers' incomes and create a huge increase in the number of long-term rural jobs. Local ownership is the key to rural prosperity. A typical biorefinery is only a fraction the size of a petroleum refinery and can be built in one year at an average cost of $75 million. America's future agriculture/energy system could be composed of thousands of locally owned biorefineries. We could move Middle-East terrorism fueling dollars to rural America.

It has been found that the optimum size ethanol plant employs about 40 people and produces about 40 million gallons a year. Thus, 3500 biorefineries of this size could meet the annual requirement of 140 billion gallons of fuel. Wind farming has become a growth industry. Farmers are earning thousands of dollars from power companies by leasing small parcels of their farms for wind turbines, on land that would otherwise earn them only hundreds if planted in crops. A University of Minnesota study found that grasses like miscanthus and switchgrass could be grown on marginal farmland without fertilizers and pesticides and produce 51% more energy per acre than corn grown on fertile land. A DOE study found that 48 million acres of switchgrass could supply 94% of U.S. transportation needs. Today the enzymes we use to convert cellulose to ethanol are expensive. Termites have enzymes that convert cellulose to sugar, which as any moonshiner knows can be converted to ethanol. DOE is working on bioengineering termite DNA to produce synthetic termite enzymes, which could cheaply convert cellulose to sugar. If DOE is successful, we will soon be able to run our cars on bug juice!

About 15 percent of the land in North America is unsuitable for food farming, but would be quite suitable for growing switchgrass. As previously noted, if only 50 to 60 million acres of this land were planted in switchgrass, we could replace every single gallon of gasoline consumed in the United States with a gallon of cheap, domestically produced, environmentally friendly cellulosic ethanol. Biodiesel CO_2 emissions are 78% lower than regular petroleum diesel; cellulosic ethanol emissions are 68% lower than regular gasoline. It has been estimated that substituting cellulosic ethanol for gasoline would slash greenhouse gas emissions by up to 90 percent.

Currently, corn yields 400 gallons of ethanol per acre; a new Monsanto variety is projected to yield 750 gallons per acre. Switchgrass currently yields 1500 gallons of ethanol per acre, it is projected to yield 2000 gallons in the near future, ultimately, is expected to yield about 2700 gallons per acre. Sugarcane is yielding 3000 gallons per acre in Brazil. Brazil is the spearhead and model for achieving independence from Middle-East tyranny. Between 2003 and 2006, Brazil gained independence from imported oil and saved $50 billion. In Brazil, each dollar spent on ethanol generates twenty times more local jobs than the same dollar spent on gasoline. Over a million new jobs have been created by Brazil's ethanol economy, and its still growing like crabgrass. Cosan is producing 3000 gallons of ethanol per acre of sugarcane. Brazil has had an almost linear decrease in the cost of producing ethanol over the past 30 years, despite minimal technology advances. They produce 50% of their fuel using only 1% of their arable land.

Alan Greenspan and venture capitalist Vinod Khosla say the energy future belongs to cellulosic ethanol made from switchgrass, powering plug-in hybrid vehicles. University of California at Berkley's Daniel Kammen: says "If the current U. S. vehicle fleet were replaced overnight with PHEVs, oil consumption would decrease 70 to 90 percent, eliminating the need for oil imports---." Kammen said we need to switch from corn-based ethanol to cellulosic ethanol, which can be made from myriad materials such as switchgrass, wood chips, stover (corncobs, cornstalks), etc. Biofuels will be much cheaper than gasoline and diesel. It has been estimated that biorefineries could produce gasoline or diesel equivalent fuels for between 60 cents and 90 cents per gallon. Khosla said: "Imagine the mid-west returned to its original prairie state, producing animal feed proteins (that's what the prairie grasses did before the row crops took over), increasing biodiversity, increasing farm income, creating rural jobs, reducing greenhouse gases and global warming, and producing ethanol or other fuels while reducing our dependence on terrorism supporting oil imports."

Khosla says we are embarking on what he calls a "biohol trajectory." Like Moore's Law, this theory predicts a steady improvement in automobile technology and a steady increase in ethanol yield per acre. This will involve cellulosic ethanol, high energy butanol, and better new biofuels with even higher energy densities. We will have engines optimized for biofuels, better batteries, plug-in hybrids, and lighter carbon composite cars. New high-compression, turbo-charged engines optimized for biofuels, will achieve efficiencies of 50%, compared to today's gas engine efficiencies of 20%. Khosla is putting his money on improving the technology of the internal-combustion engine and cellulosic ethanol processing. To wit: EcoMotors is developing a two-cylinder, two-stroke diesel engine that is 30% lighter, 25% the size, and achieves 50% better fuel economy than a conventional turbodiesel. He says improved cellulosic ethanol processing technology will make biofuels cheaper than gasoline within five years.

A few historical notes. The first hybrid car was the 1915 Owen Magnetic, it had a gas engine and an electric generator. It had a 24-volt electrical system and no transmission; only a rheostat. Owens were expensive and sophisticated. Enrico Caruso and John McCormack were proud owners. The Owen Magnetic tanked in 1921. Rudolph Diesel's original car burned peanut oil. In 1908, Henry Ford's Model T used 200-proof ethanol (moonshine) or gasoline as a fuel. Henry's Model T got 25 mpg; today, the average American car gets 21 mpg. Recently, a 1921 Model T beat a 2003 Hummer H2 in a hill climb in Evansville, Indiana. Ford was a strong supporter of homegrown renewable fuel. Ethanol was used to fuel cars well into the 1920s and 1930s. In the 1920s, Standard Oil marketed a 25% ethanol/gasoline mix in the Baltimore area. In 1938, an ethanol plant in Atchison, Kansas was producing 18 million gallons of ethanol a year to supply 2000 service stations in the Midwest. After World War II, the government lost interest in ethanol production; petroleum and natural gas were cheap and plentiful. Wartime distilleries were dismantled or converted to alcoholic beverage plants.

The biofuel bandwagon is beginning to roll, the really heavy hitters are jumping on board: DuPont, Exxon, BP, Chevron, W.R. Grace, Honeywell, etc. DuPont is working with Diversa to develop enzymes that break down cellulose into sugars. It is teaming with DOE's National Renewable Energy Laboratory (NREL) to develop new fermentation processes to produce cellulosic ethanol. DuPont teamed with Brion to upgrade one of their biorefineries to produce ethanol from corn and stover. DuPont says: "we are close to making cellulosic ethanol a reality." DuPont has teamed up with oil giant BP to make butanol from biomass. Butanol and ethanol are both alcohols, but butanol has a higher energy content and blends more easily and in higher concentrations with gasoline. Biobutanol can be dumped directly in the tank without retrofitting the car. It is also compatible with existing pumps and pipelines. DuPont, BP, and British Sugar are converting an ethanol fermentation facility to make butanol. They plan to develop a biological catalyst to ferment biomass-derived sugar to butanol. Chevron is working with DOE's NREL to produce ethanol and other fuels from forestry and agricultural waste. Amyris Biotechnologies are adding genes to microbes to make a diesel and gasoline-like product that they say is better than ethanol or biodiesel. America enjoys a unique combination of creative scientists, entrepreneurs, venture capitalists and private equity funds that will launch economic booms in industry and agriculture comparable to the internet boom. Jim Woolsey quipped: "You can't stand on a street corner in Silicon Valley today without some venture capitalists throwing money at you for an ethanol plant."

MIT's Plasma Science and Fusion Center and the Pacific Northwest National Laboratory (PNNL) have developed a new incineration process that uses lightning-like plasma arcs to produce ethanol and methanol from garbage. The process yields six times as much energy as it consumes. The process is being commercialized by Integrated Environmental Technologies (IET). There is enough energy in municipal waste to replace 25% of U.S. gasoline consumption. Conoco and Tyson Foods are teaming up to produce diesel from leftover poultry, pork, and beef fat. Panda Ethanol Inc. is building a plant that will extract methane from one billion pounds of manure, the output of about 500,000 cows, to annually produce 100 million gallons of methanol; the plant begins operations in 2007. Panda plans to build plants in Kansas and Colorado.

In the beginning, over ninety percent of the American people were farmers; today only two percent are. Between 1900 and 1920, America had a “Golden Age of Agriculture.” In 1909, Congress passed the Enlarged Homestead Act, which gave settlers 320 acres of land in exchange for five years residency and improving the land. During this period, the average gross income of farms more than doubled and the value of the farms more than tripled. Today, we stand on the threshold of a new renaissance of American agriculture. Biofuels will usher in a new “Golden Age of Agriculture.” The agriculture industry is about to expand to the point that Jefferson’s old dream finally will be realized. America will become a nation of yeoman farmers – 21st century Centurions!

ULTRALITE

In 1991, GM built the Ultralite to demonstrate the advantages of lightweight composites and low-drag aerodynamics. The Ultralite had a carbon-fiber composite body that weighed 420 pounds (190-kg); the body consisted of only six parts: a floor tub, a right and left body half, a right and left door, and a rear panel. It was designed and built by GM's Tech Center in Warren, MI and Burt Rutan's Scaled Composites in Mojave, CA. It was all done by 50 people in 100 days at a cost of between $4 and $6 million.

The Ultralite weighed 1400 pounds, could carry four passengers, had air conditioning and a 4-speed automatic transmission. The engine and transmission were in a quickly removable "powerpod." It had an aerodynamic drag coefficient of only 0.192, about half that of the average four passenger car. It was powered by an aluminum 1500cc, direct injection 2-stroke 111 HP engine. It would go from zero to 60mph in 7.8 seconds, had a top speed of 135mph, and got 100 miles per gallon. The Rocky Mountain Institute estimated that as a hybrid it would get 200 miles per gallon.

Studies conducted at that time showed that composite bodies like the Ultralite had the potential to be up to 67% lighter than conventional a steel unibody of equivalent size and safety. Stamping and welding steel body parts is much faster than assembling composite parts; but this is partially offset by a huge decrease in the number of parts and assembly steps. Steel cars require massive investments in tools and equipment; each new model year is a multi-billion dollar, bet-the-company gamble.

The Ultralite prototype carbon-fiber monocoque body was not optimized for mass or manufacturability; it weighed 190-kg with closures. Swiss designers at Oerlikon estimated they could build a body as big and safe as Ultralite that would weigh approximately 72-kg; greatly reducing the cost of the composite materials. Glass is one of the heaviest components of a car; it will be replaced by new polycarbonates similar to those used to cover headlights. The polycarbonate will be coated with a thin layer of glass; it will be 50 percent lighter than glass. DARPA is working on combining carbon fibers and carbon nanotubes to produce a material ten times as strong as existing composites. It will give us cars "light as a feather; strong as a tank," as my brochures described the March Hare and ECCO.

By greatly reducing the weight of the body, the size and weight of the skateboard "powerpod" (engine, electric motors, generator, batteries, ultracapacitor, drive-by-wire components, etc.) is greatly reduced. The aircraft industry found that by designing for optimized composite manufacturing techniques, the composite mass fraction was increased from 28% to 95%, while the total production cost was reduced by 56%. 50% (by weight) of Boeing's 787 fuselage is composite, compared with 12% on the 777. Due to composite construction, the 787 uses 80% fewer fasteners than the 777. Researchers in the mid-1990s estimated that an advanced composite body could be built cheaper than a similarly sized conventional steel unibody. They estimated that an advanced composite monocoque car could break even in manufacturing cost with a steel unibody car at volumes of 100,000 vehicles per year.

Bill Ford, CEO of Ford Motors said the 100-year reign of the gas-powered internal combustion engine (ICE) is ending. An electric-powered transportation system will replace the mechanical system. Polymer composites have already displaced metal in the pleasure boat industry; they are rapidly taking over aircraft industry; the automotive industry finally appears to be on the threshold of taking the giant step. The Navy's nuclear carriers and submarines are hybrid electrics; the nuclear reactors generate steam to drive the electric generators. The Navy's next generation surface fleet will be all-electric ships equipped with hypervelocity (Mach 8-10) electric guns. Electric hypercars will create a global shift in industry comparable to that created by microchips. Hypercars will be like computers with wheels; Silicon Valley will play the principal role in their design and development.

Hybrid power is not a disruptive technology requiring massive capital investment. The Toyota plant that assembles the Prius also assembles several other models. The Prius only requires four new parts and eleven additional procedures (out of 200). Prius production costs have fallen more than 50% in the past seven years. Toyota expects costs to fall another 50% on the 2008 model.

Toyota's Prius's are series/parallel hybrids, powered by the gasoline engine and/or the electric motor. A power-split device links the engine, generator, electric motor and the drive shaft, which drives the wheels. It is a complex and expensive system; it will be interesting to see if Toyota sticks with it. In a series hybrid, such as GM's Volt, the wheels are powered solely by the electric motor, the internal-combustion engine only drives the generator to produce electricity to recharge the battery pack. GM's Volt series hybrid system is simpler and cheaper to build than the Toyota Prius series/parallel system. The Volt functions as a true electric car with an ICE backup. However, the power and acceleration of a series hybrid is inferior to that of a series/ parallel. This disadvantage can be overcome by installing an ultracapacitor to give the series hybrid a boost in power and acceleration, when needed. Hybrid technology is compatible with many fuel-saving technologies being developed for gasoline or diesel engines, such as cylinder de-activation, variable valve timing, direct fuel injection, advanced turbochargers, smaller engines, and friction reduction.

Batteries can be designed for power density, which enables the electric motor to accelerate faster and improves fuel efficiency by partially displacing the gas engine during low speed acceleration. Conversely, the battery can be designed for energy density, which increases the range but reduces acceleration. Again, as in the case of the series hybrid versus the series/parallel, this disadvantage can be overcome by installing an ultracapacitor.

Currently, lithium-based batteries are more expensive than nickel-based batteries, but with volume production, they are expected to become cheaper. The cost of lithium batteries fell by more 86% in the ten year period 1994 to 2003. Lithium batteries have twice the energy density and three times the power density of nickel batteries. Lithium batteries are about 50% lighter and smaller than nickel batteries of comparable power and energy density. The problem is that current lithium batteries use lithium cobalt oxide in the cathode, which under certain conditions, can be explosive. This can be resolved by replacing the cobalt oxide with manganese oxide, or various combinations, such as nickel cobalt and manganese or nickel, cobalt and aluminum. Other possible approaches include lithium polymers and lithium sulfur, which have the potential of greater energy density. Iron phosphate may be the most promising new cathode material. A123 is using it in the battery it is developing for GM. Iron phosphate is inexpensive, and because the bonds between the iron, phosphate, and oxygen are much stronger than those between cobalt and oxygen atoms, the oxygen is much harder to detach when overcharged; thus, when it fails, it does not overheat.

Electronic components are more precise, reliable, efficient, smaller, lighter, and cheaper than mechanical components. Electrical systems do not break down as frequently as mechanical systems. Electric motors are smaller, lighter, and up to five or six times as efficient as internal combustion engines. According to industry experts, 80% to 90% of automotive innovation is based on electronics. They predict electric/electronic systems will account for 70% to 80% of the cost of hybrid vehicles in the next five to ten years.

PHEVs equipped with 300-volt batteries will lead to full electrification of the automobile; mechanical systems will be replaced by drive-by-wire electric systems. New batteries will last the life of the car and electric motors are good for a million miles. Since regenerative braking helps slow the car down, brakes get used less and last longer. In the aerospace industry, all advanced aircraft are fly-by-wire. Brake-by-wire and steer-by-wire will replace hydraulic pumps and reservoirs, fluids and hoses with electronic components that are smaller, lighter, safer, more responsive, and more reliable. Shift-by-wire will replace conventional transmissions with electronically controlled systems. Throttle cables and belts that connect air conditioners, oil pumps, water pumps, cooling fans, etc., will be replaced. Suspension-by-wire will permit total control of pitch and roll when braking or cornering; the car will lean into curves like a motorcycle. The Supercar could take the form of a GM Ultralite-type body mounted on a GM Autonomy-type skateboard chassis. As with the Autonomy, the body could be plugged into the chassis, permitting the owner to enjoy a variety of body types, such as: sedan, sportscar, classic replicar, truck, etc. GM's last skateboard, the Sequel, used steer-by-wire and brake-by-wire.

Buses, trucks and military vehicles stand to benefit greatly from hybrid technology. Globally, there are over 1.5 million buses on the roads; using about five percent of the fuel consumed. Buses travel 50,000 to 100,000 miles per year and average seven to nine miles per gallon. An NREL study of hybrid buses operating in NYC found they delivered 45% better fuel efficiency than diesel buses and 100% better than natural gas buses. Globally, there are about 46 million medium and heavy-duty trucks on the road. As with buses, they travel 50,000 to 100,000 miles per year and average four to ten miles per gallon. Hybrid technology is expected to improve fuel efficiency 30% to 70%. DOD is America's largest oil consumer; a Pentagon study found that fuel accounted for about 70% of the cargo tonnage of military convoys. Abrams tanks burn one gallon of fuel per mile. Until recently cancelled, the Army was developing and building a new family of hybrid-electric Manned Ground Vehicles (MGVs). All eight variants would have a common chassis. They would have been series hybrid systems in which the engine would recharge the energy storage system. The MGV family would have included a tank with a lightweight 120mm cannon, a self-propelled howitzer with a 155mm cannon, and an Infantry Carrier Vehicle with a 30mm cannon. All MGV gun systems are fully automatic.

The widespread adoption of PHEVs, which will usually be charged during off-peak hours at night, will help utilities achieve level loading, which improves efficiency and lowers the unit cost electricity. This is because nuclear plants and some coal-fired plants, generate a steady flow of power regardless of demand. Nationwide adoption of PHEVs will launch a golden age for the electric utility industry by opening it up to the vast new world of road transportation. This new river of revenue will enable the industry to upgrade the grid and invest in cleaner and more efficient power plants. DOE's Pacific Northwest National Laboratory estimated that widespread adoption of PHEVs will not create a need for new electric power plants until over 84% of the country's 250 million cars are plug-ins. Most U.S. cars are parked more than 20 hours a day. If PHEVs were left plugged in after their batteries were charged, they could provide stabilization backup to the national grid, this is called vehicle-to-grid (V2G). Owners of solar panels and windmills already sell their excess electricity back to the power companies. America's utilities spend about $10 billion a year on grid stabilization. Studies have shown that if only 3% of the nation's cars were PHEVs that were plugged into the grid, they alone would be able to handle the grid stabilization problem. Old PHEV battery packs could be used as a home emergency power source and as V2G backup. Volkswagen estimated that one million V2G PHEVs could generate electricity equivalent to 20 average-sized power plants. DOE estimated that widespread adoption of V2G technology could increase wind power usage by a factor of three. Most interesting of all, it is estimated that each V2G PHEV might earn its owner as much as $3000 per year.

In their report, *Oil and Security*, George P. Schultz and R. James Woolsey called the potential of these new technologies stunning: "For example, a 50-mpg hybrid gasoline/electric vehicle, on the road today, if constructed from carbon composites would achieve around 100 mpg. If it were to operate on 85 percent cellulosic ethanol or a similar proportion of biodiesel or renewable diesel fuel, it would be achieving hundreds of miles per gallon of petroleum-derived fuel. If it were a plug-in version operating on upgraded lithium batteries so that 20-30 mile trips could be undertaken on its overnight charge before it began utilizing liquid fuel at all, it could be obtaining in the range of 1000 mpg (of petroleum)."

ULTRACAPACITORS

Compact energy storage is the principal roadblock to an all electric-powered transportation system; engineers and scientists all over the world are working on the problem. Ultracapacitors may be the answer: ten year plus lifetime, indifference to temperature, high immunity to shock and vibration, contain no hazardous materials, recharge in minutes. Capacitors store energy as an electrical field, making them more efficient than standard batteries, which get their energy from chemical reactions. Storage capacity is proportional to the surface area of the electrodes. MIT's Laboratory for Electromagnetic and Electronic Systems is experimenting with vertically aligned nanotubes, only several atomic diameters wide, to provide an enormous increase in surface area and storage capacity. EEStor, Inc., a Cedar Park, Texas company is developing an ultracapacitor that they say could be recharged in five minutes and provide enough energy to drive 500 miles. For PHEVs greater energy density is more important than power density, because energy density provides longer electric-only driving range. An ultracapacitor could be installed to provide bursts of higher power. The science of capacitors has been understood for over a century. Storage capacity is proportional to the surface area of the electrodes. Electrons are compressed around the nucleus. Think of a swiss cheese sandwich. The bread holds the opposite electrical charges, the electrodes, the cheese separates the slices of bread, and keeps the charges from cancelling each other out. Storage capacity is proportional to the surface area of the bread and cheese, the nooks and crannies of a rough slice of bread, like Thomas's English Muffins, can hold more mayonnaise than a smooth slice.

United States Patent number 7,033,406 was issued to Richard D. Weir and Carl W. Nelson of EEStor, Inc on 4/25/06 for an "electrical-energy-storage unit (EESU) utilizing ceramic and integrated-circuit technologies for replacement of electrochemical batteries." The abstract says the EESU "has as a basis material a high-permittivity composition-modified barium titanate ceramic powder. This powder is double coated with the first coating being aluminum oxide and the second coating calcium magnesium aluminosilicate glass." EEStor says their EESU ultracapacitor can pack ten times the energy of a PbA battery of the same weight and cost half as much. They claim it has double the energy density of a Li-ion battery at one-eighth the cost. It can be recharged in under five minutes versus hours for PbA or Li-ion batteries. PbA batteries are good for a maximum 500 to 700 recharge cycles; EESUs can be recycled a million times with no degradation. They do not overheat, they are not explosive, corrosive, or hazardous, unlike PbA or Li-ion electrochemical batteries. The EESU is designed for millions of 100% charge/discharge cycles; able to operate in temperature ranges of –20 to +60 degrees Celsius. The described EESU weighed about 336 pounds, measured 13.5"x13.5"x11"(a little over a cubic foot), and had a storage capacity of about 52 kWh. It is projected to power the Zenn for the equivalent of 45 cents a gallon.

EEStor plans to introduce the EESUs in 2010. Ian Clifford, CEO of Zenn Motors had planned to produce an EESU-powered electric car able to drive from Toronto to Montreal on $9 worth of electricity. As for performance, Clifford said: "a four-passenger sedan will drive like a Ferrari." Some are calling the EESU "the holy grail of battery technology." It could become the Google of the clean tech world. EEStor is being backed by venture capital giant Kleiner Perkins and Byers, the folks who brought us Google, Amazon.com, and Netscape. Such ultracapacitors could store electric power generated by voltaic cells and windmills; provide backup power units for residential, commercial, and industrial sites.

MIT researchers have proposed a new engine design employing a higher compression ratio and a turbocharger to increase power by 250%. High compression ratios cause spontaneous combustion, called knocking. MIT solves the knocking problem by precise injection of ethanol, which has a higher octane rating than gasoline. The ethanol cools the mix and increases octane rating to about 130 – as good as high-performance racing fuel. The MIT engine is projected to be 30% more efficient than a conventional engine.

America stands on the threshold of complete energy independence. Solving America's petroleum problem is not rocket science or even quantum mechanics. We could build enough refineries to convert our coal to diesel or methanol, or to convert our oil shale to gasoline; that would solve the problem. We could put PHEVs with 30 to 40 mile electric ranges into mass production; that would solve the problem. We could plant about 50 million acres of switchgrass and build the required ethanol refineries to produce annually approximately 140 billion gallons of ethanol; that would solve the problem. We could extract the residual oil from our 400,000 capped oil wells; that would solve the problem. We could capture the CO_2 emitted by our powerplants and factories to fuel algae bioreactors and be awash in an ocean of biofuel; that would solve the problem in spades.

America needs to end the "strain at a gnat and swallow a camel" approach to energy conservation. It is noble to encourage people to buy energy efficient light bulbs, TVs, and refrigerators, but it is absurd in the face of the horrendous energy wasted lighting up places like Las Vegas, Times Square, and big cities all over the country. People buy energy efficient TVs, then leave them running in empty rooms all over the house. A major portion of the energy problem could be solved by ending such wasteful practices. We need to start applying common sense and logic to the problem.

Widespread adoption of these new energy technologies would have momentous geo-political consequences. Economic well being is closely linked to transportation; which could be almost completely decoupled from petroleum. This would bring an economic boom for oil-importing countries and an economic catastrophe for oil-exporting countries. America does not have much time. In a few years a billion Chinese and a billion Indians will be lining up to buy cars. The U.S. can drain Alaska dry and dot our coast lines with oil rigs, but we can't drill our way out of this global dilemma. Implementing these new technologies would trigger a renaissance in industry, technology, and agriculture and set in motion a nationwide wave of prosperity that would last for decades. Alternative energy is the greatest economic opportunity of the 21st century; some venture capitalists say it is going to be an order of magnitude greater than the Internet!

CARS FOR THE MASSES

"Fat men cannot run as fast as thin men, but we build most of our vehicles as though deadweight fat increased speed --- I cannot imagine where the delusion that weight means strength came from ---." Henry Ford

Except for a few mavericks, like Henry Ford, the car industry always preferred to build big cars because that's where they made the big profits, ala Chevrolet Caprice/Cadillac DeVille. A tectonic shift is currently underway in the auto industry; a renaissance of Henry Ford's philosophy. Manufacturers in Asia are gearing up to produce low-priced cars for the masses. India's Tata Motors launched a $2000 car in 2009. The Tata is a four door car with a 33 horsepower engine and a top speed of 80 mph. Every major Asian carmaker has a 21st century Model T project underway. GM's Korean subsidiary Daewoo is designing a $7000 car. Toyota is tooling up to produce an under $7000 car in 2009. Toyota is pursuing an all-out cost-cutting campaign in design, materials, and manufacturing to be applied across their entire line. They plan to cut the cost of building the Prius by 50% and are gearing up to produce a low-cost PHEV. Renault's Romanian built $7000 Logan showed the way by eliminating prototypes and moving directly from digital mockup to production. Boeing is doing the same thing with the 787 Dreamliner.

Batteries are at the heart of every electric or hybrid car of the future. Almost all batteries made in the world are made in Asia; primarily in Japan, South Korea, and China. American battery companies are required to turn to Asia for mass production. China, Japan, and South Korea provide long-term, multi-billion dollar support to their battery companies. A modern lithium ion battery plant, like those in Japan, are highly automated operations; they are very much like chip manufacturing plants. In chip manufacturing plants, the year-to-year advances in capacity, quality, and price, come from constant manufacturing improvement. American companies that in the 1980s decided to be "fab-less semiconductor companies," outsourced their chip fabrication to Asia. Those companies never again made chips. American companies, like Intel, that stayed in the game, and are the best and most competitive chipmakers in the world. As things stand now, if the U.S. and Europe plan to mass produce PHEVs, they will have to buy the batteries in Asia.

Globalization of manufacturing is accelerating. Asian companies send their parts to China for assembly. High-tech components from Japan are shipped into China for assembly into the final product with cheap labor. Asia is gearing up to revolutionize the auto industry. Asia appears to be positioning itself to become the world's supplier of low-cost electric cars. In the 1950s, 60s, and 70s, Volkswagen flooded the world with practical, reliable, low-cost Beetles. It appears the world is about to be flooded again with practical, reliable, low-cost Asian PHEVs/EVs.

With a population of 300 million, America has around 250 million cars. With a population of over one billion apiece, China and India are the most rapidly industrializing countries in the world. It is only a question of time before each has over 400 million cars. Because of their lower wages, they are going to be building the world's low-priced cars. This may force American and European carmakers to build the best cars they can build, instead of building what they can get away with. There was a time in America when we had companies like Duesenberg and Leland that tried to build the best cars they could build. At the moment, America and Europe enjoy a technological edge over China and India. No one can say how long this will last, but we had better try to gear up to "make hay while the sun shines."

SUPERCARS

One way we can try, is to start designing the Supercars of the future right now. In the body of this report, we have outlined many of the characteristics such a car will probably incorporate. We know some of the technologies that need to be pushed: batteries, capacitors, composites, drive-by-wire, multi-fueled engines, electric motors, polycarbonate glass, manufacturing techniques, etc.

After a century of automotive engineering, only 15% to 20% of a car's fuel energy reaches the wheels, 95% of that moves the car, only about 5% moves the driver. Steel cars are so heavy large engines are required to accelerate the mass. Engines are ten times larger than needed for most driving. Only about 16% of an engines power is used on the highway; severalfold less in the city; only about 40hp is needed to sustain a 65mph cruising speed. The National Academy of Science recently estimated that up to two billion gallons of gasoline and diesel fuel could be saved in the U.S. each year by reducing the rolling resistance of automobile tires by 10%. That's the equivalent of taking four million cars and light trucks off the road. The Academy said the safety consequences of such changes "are probably undetectable."

Advanced composites are being increasingly used in the aerospace, boating, sports equipment, and building materials industries. Their increasing use and falling costs are making them all the more viable for the automobile industry. Composite bodies can dramatically reduce weight while increasing strength and improving durability. Lighter vehicles require less power to accelerate, climb hills, cruise, and stop. This lower power requirement yields smaller, lighter, and lower-cost drivetrains.

America's long-term love affair with the "safety" of large and heavy vehicles sometimes appears to be hopeless and eternal. Tens of millions of 5000 to 6000 pound gas-guzzling SUVs clutter our highways, often transporting a cargo consisting of a 5 foot 2 little woman peering over a steering wheel. This is nothing new. For most of the 20th century Americans were told heavy cars "hold the road better". The racing industry should have demonstrated the idiocy of this epigram. Fundamental physics illustrate the role of mass in colliding bodies. The average American passenger sedan weighs about 3500 pounds. An ultralight/ultrastrong carbon composite PHEV/EV with an integral roll cage, should weigh about 1500 pounds. The formula for kinetic energy is $1/2\ MV^2$.

> Two 3500 pound cars colliding head-on at 30mph, generate 13,552,000 lbft/sec2 energy.
>
> Two 1500 pound cars colliding head-on at 30mph, generate 5,808,000 lbft/sec2 energy.
>
> Clearly, highways filled with ultrastrong 1500 pound vehicles would be significantly safer.

Ultralight construction brings into play the mass decompounding factor, in which components not only shrink, but may disappear. Saving one kilogram of mass might save up to five kilograms indirectly. An ultralight platform needs no power steering or power brakes. A PHEV needs no transmission, flywheel, clutch, starter, alternator, driveshaft, universal joints, differentials, and perhaps axles. Today, automakers are hybridizing heavy production platforms, which means that mass, complexity, and costs are compounded. Whereas, hybridizing an ultralight and ultra-low-drag platform yields striking advantages in mass decompounding.

A correctly designed composite structure can absorb up to five times as much energy per pound as steel. Rocky Mountain Institute (RMI) said the strength of a monocoque structure can be demonstrated by trying to eat an Atlantic lobster without tools. Carbon-fiber composites are even stronger; they can absorb an order of magnitude more crash energy than steel or aluminum. Millions have watched carbon composite Indy 500 racecars crash into the walls at 200mph; usually, the driver climbs out and walks away. RMI described their Hypercar as "people, cushioned by foam, surrounded by a Macadamia-nut shell, wrapped in thick bubble-pack."

Composite manufacturing offers the potential for 3 to 10 times lower product cycle time, tooling and equipment investment, assembly space and effort, and body parts count. Future cars could have 2 to 3 times lower curb mass, 2 to 6 times lower aerodynamic drag, 3 to 5 times lower rolling resistance, and 3 to 10 times lower accessories load than today's conventional car.

In World War I, Dr. W.W. Christmas planned to kidnap the Kaiser with a squadron of steel biplanes. An efficient car cannot be made out of steel for the same reason that an efficient aircraft cannot. Thanks to the laws of physics, when cars are designed less like tanks and more like aircraft, wonderful things start to happen. From the beginning, the auto industry has focused its attention on improving the efficiency of the engine and drivetrain. The Hypercar, Inc. concept attempted to improve the efficiency by replacing the stamped steel body with a body made of advanced composites, such as carbon fibers, Kevlar (polyaramid), glass, and other ultrastrong fibers embedded in moldable plastic.

Carbon fibers are black, shiny, stiff filaments finer than a human hair; one-fourth as dense as steel but stiffer and stronger. Properly aligned and interwoven to match the loadpath and distribute stresses and embedded in a strong polymer matrix, it forms a composite with the same strength and stiffness as steel, but two to three times lighter. Combinations of myriad fibers and plastics offer design flexibility to exactly match the properties needed in a given part. Only about 15% of the cost of a typical steel part is for the steel itself; 85% pays for pounding, welding, and finishing. Composites emerge from the mold in their final shape. Very large and complex parts can be molded in a single piece. A composite body needs only five to twenty parts; a steel unibody needs two hundred to four hundred. Each of those hundreds of steel parts needs an average of four tool-steel dies to pound it into shape; each die costs an average of $1 million. Polymer composite parts are molded to their final shape in a single step, using plastic molds that cut tooling costs by 90%.

The lightweight parts can be lifted without a hoist, they snap together precisely without rework, and are joined using superglues instead of by hundreds of robot welding machines. Assembly costs are reduced by 90%. Painting steel body parts accounts for 25% to 50% of the cost of the finished part. The final color of a composite part is sprayed into the mold before the composite materials are laid up in the mold. At every level of manufacturing, the differences between using steel and composites are enormous. For a conventional new car model, a thousand engineers will spend a year designing and another year building a billion dollars worth of car-sized steel dies; enough to fill a football field. The "bet-the-company" costs take years to recover and require huge production runs that last for longer than market projections can forecast.

Hypercar, Inc. has a patent-pending process called Advanced Volume Automotive Composite Solution (AVACS). The process automatically and rapidly lays carbon and other fibers in the desired orientation on flat "tailored blanks", then thermopresses the blanks to their final shape on a moving line. They use long discontinuous fiber (LDF) carbon. LDF is cheaper and easier to work than the continuous fiber used in aerospace. LDF can stretch during molding, but the fibers are long enough to maintain near continuous fiber levels of stiffness. Hypercars goal is to attain 80% of the performance of hand-layup aerospace composites at 20% of their cost. The Hypercar strategy utilizes small "Skunk Works" design teams, low production runs, low break-even volume per model, rapid experimentation and model diversification.

Fantastic new materials may change everything. Dr. Richard E. Smalley, Professor of Chemistry and Physics, Center for Nanoscale Technology at Rice University thinks it may be nanotubes. He said nanotubes have myriad and marvelous thermal and electrical properties. They are roughly 30 to 100 times stronger than steel and 10 to 100 times stronger than carbon fiber. Another revolutionary material is called metallic glass or amorphous metal. Originally developed by a group at Caltech/Pasadena, working under material science professor, Dr. William Johnson. It's a new class of material six times as strong as steel, yet it bounces like a rubber ball. It can be formed like a plastic. To make a car fender, instead of making sheet metal, then cutting, stamping, drilling, etc., the fender can be injection molded in one piece. Amorphous metal can be made into foam panels that are 99% air but 100 times stronger than polystyrene. A sandwich made of two thin sheets of amorphous metal enclosing amorphous foam, would be light, strong, rustproof, insulating, sound dampening, bug-proof, and like armor. Such panels could be used to construct buildings, ships, aircraft, and car bodies.

Amorphous metals are superior to conventional metals and plastics. They may be the structural material of the future.

RMI's 1996 Hypercar design study was for a series hybrid powered by a Stirling engine, a 48kW permanent magnet motor, and a 300 volt PbA battery. It had a curb weight of 1287 pounds, could carry 4-5 passengers, and had the same Cd as GM's Ultralite and EV-1 – 0.19. It was designed to go from 0-60 in 8.6 seconds and get 110 mpg.

Hypercar, Inc. designed the Revolution, a five-passenger SUV crossover concept vehicle to demonstrate RMI technologies. It is a fuel cell powered hybrid that weighs 1800lbs, has a Cd of 0.26, 0-60mph in 8.2 seconds, drive-by-wire, 114mpg equivalent, and 500 mile range. Body stiffness and torsional rigidity are 50% higher than premium sports sedans. A 35mph crash into a wall won't damage the passenger compartment. It has 14 major parts and 62 total parts; 65% and 77% fewer parts respectively than a conventional stamped steel body. Using minimal jigs and fixtures, the major parts snap together using blade and clevis and quick-loc adhesive. The most advanced aircraft flying today, Boeing's X-45C UCAS was built by fewer than 20 people with no hard tooling. It has an aluminum substructure with an all-composite skin in three parts – the fuselage and two outer wings. All clips and brackets are bonded into the structure.

Today's average production sedan has a drag coefficient (Cd) of 0.33; GM's EV-1 had a Cd of 0.19; Ford's 1980s Probe V concept car had a Cd of 0.137. The 1935 Czech Tatra 77a had a Cd of 0.21. They were banned after the Anschluss; wild young Wehrmacht officers were crashing them in the mountain roads. Some aerodynamicists believe a Cd of about 0.08 is possible with some form of passive boundary-layer control somewhat like dimples on a golf ball. The theoretical minimum drag shape is a teardrop 2.5 – 4 times as long as its maximum diameter with a Cd of 0.03 – 0.04.

In 1991, GM built the Ultralite to demonstrate the advantages of lightweight composites and low-drag aerodynamics. The Ultralite had a carbon-fiber composite body that weighed 420 pounds (190-kg); the body consisted of only six parts: a floor tub, a right and left body half, a right and left door, and a rear panel. It weighed 1400 pounds, could carry four passengers, had air conditioning and a 4-speed automatic transmission. The engine and transmission were in a quickly removable "powerpod." It had an aerodynamic drag coefficient of only 0.192. It was powered by an aluminum 1500cc, direct injection 2-stroke 111bhp engine. It would go from zero to 60mph in 7.8 seconds, had a top speed of 135mph, and got 100 miles per gallon. The Rocky Mountain Institute estimated that as a hybrid it would get 200 miles per gallon. GM's Ultralite was built by 50 people in 100 days for between $4 million and $6 million; GM's EV-1 was productionized in one year. Hypercars are well suited to Skunk Works type projects; surging from bench to pilot to full production.

A123Systems, an MIT spin-off, has developed a lithium-ion phosphate battery pack that can be installed in the spare tire well of the Prius in two hours by trained mechanics. It converts the HEV Prius into a PHEV with a 40-mile electric only range. The module recharges overnight from a 120-volt outlet. A123Systems was working with GM on the Volt project. A123 is working on batteries about the size of a carton of cigarettes, that weigh about ten pounds each, and can be charged to 90% capacity in 5 minutes and fully charged in 15 minutes.

GM's Chevy Volt is a series PHEV; the front wheels are driven by a 161-hp (120-kW) electric motor, through a planetary reduction gear. A lithium-ion battery pack, located along the car's central tunnel, powers the motor for 40 miles. When the battery pack charge falls below 30%, a turbocharged 1000cc three cylinder 85-hp engine kicks in to power a 53-kW generator. Current from the generator is routed to the battery pack. The engine shuts down when the batteries are recharged to 80%. Diesel-electric locomotives, developed by GM in the 1930s, work the same way. Twin six gallon flex-fuel tanks give the car a 640 mile range. The Volt's current configuration weighs 3200 pounds and accelerates 0 – 60 in 8.5 seconds. With an ultracapacitor, which every series hybrid should have, the 0 – 60 time could be greatly reduced.

GM is building a $43 million plant south of Detroit to assemble batteries from South Korea's LG Chem Ltd. into 70,000 packs per year. Each pack will have 220 cells and is currently estimated to cost $8000. The packs will be used in the Volt and other GM PHEVs. GM will design, develop and manufacture its own electric motors at its Baltimore transmission factory. GM plans to build the Volt's gasoline engine in Flint, Michigan; its body and chassis will be built at its Hamtramck plant in Detroit. Volt production is scheduled to begin in November 2010.

At the Shanghai Auto Show, GM unveiled a second propulsion system for the Volt. It's a hydrogen fuel cell variant that uses GM's new fifth-generation fuel cell system as its primary power source. The E-Flex system combines a new 80kW fuel cell stack with a 8kWh (50kW peak power) lithium-ion battery to yield a range of 300 miles.

GM plans to produce a European version of the Volt – the Opel Flextreme. The Flextreme will use a four cylinder, 1.3 liter CDTI turbo-diesel engine, that produces 53 kW peak power. The nanophosphate lithium-ion battery pack has a peak power of 136kW and a voltage of 320V to 350V. It can be recharged in three hours from a standard 220V socket. Flextreme will accelerate from 0-60mph in 9.5 seconds, have a top speed of 100mph, an all-electric range of 34 miles, and a total range of 444 miles. It will come equipped with two Segways packed below the cargo floor. GM plans to start road testing in 2008 and start production by year-end 2010.

At the 2009 Frankfurt Auto Show, tire maker and restaurant critic Michelin unveiled what could become a three-star development. Michelin showed two versions of its Active wheel, one with an integrated electric motor, the other with a motor and a suspension system. The motor in each Active wheel 30 kilowatts (kW) of continuous power, and 60kW of peak power. Two Active wheels on a car generate the equivalent of 80 horsepower; four add up to 160 horsepower. The motors act as a regenerative braking system. The integrated suspension system is active and can respond to body roll and pitch with 145mm of travel. If equipped with suspension-by-wire, the car will be able to lean into curves like a motorcycle. A car equipped with a set of these wheels could eliminate the drivetrain and suspension components. Michelin Active wheels appear to be the next step down the road from British engineering company's PML Flightlink.

Several years ago, PML Flightlink, put together a high-performance PHEV version of BMW's Mini Cooper, the PML MINI QED. PML took a standard Mini One, and discarded the engine, the wheels, the disc brakes, and the gearbox. These components were replaced by four PML electric wheels, a lithium polymer battery, a large ultracapacitor, a small internal combustion engine (ICE), and an electronic management system. The car had a top speed of 150mph, a 0-60mph time of 4.5 seconds, an all-electric range of 200-250 miles, and a total range of 932 miles (1500 km).

Each wheel developed 160bhp – 640bhp total. The Mini engine that was replaced developed 100bhp and weighed twice as much as the four electric wheels. All braking was done with the wheel motors acting as generators, returning the energy back to the battery and ultracapacitor. As the battery power level was reduced, the ICE/generator kicked in to recharge the battery and ultracapacitor. In this hybrid mode the car goets 80mpg. It could be plugged into the power grid when you got home. The PML in-wheel drive technology is adaptable to any vehicle chassis. It eliminates brakes, gears, and the mechanical drive train; it creates more space inside the car. Mitsubishi is said to have a similar system "under development"; Volvo has already acted.

Volvo's California think tank, Volvo Monitoring and Concept Center teamed up with PML Flightlink to design the Volvo ReCharge PHEV. Built on Volvo's C30 platform, it has a four-cylinder 1600cc Flexifuel engine, four electric wheelmotors, and a lithium-polymer battery pack. It has an electric-only range of 100km (62miles); full recharge takes 3 hours. It accelerates from 0-60 in 9 seconds, and has a top speed of 100mph. For a 150km (93 mile) trip, it gets 124mpg. There is no transmission, power to each wheel is controlled individually, brakes are electric. All this is controlled by a quadruple-redundant electronic control system. Michelin developed special high-efficiency tires designed for wheelmotors. The Flexifuel engine kicks in when the battery pack reaches a 30% state of charge. A button on the dash permits the driver to start the engine earlier to keep the charge at a higher level.

Another new car based on PML Flightlink, Ltd technology is the Lightning. Constructed of light and strong aluminum honeycomb, the car accelerates from zero to 60mph in under four seconds and has a top speed of 130mph. The Lightning's most interesting feature is its batteries, which the developers say have a range of 250 miles, a full recharge time of only 10 minutes, and a 12-year lifetime.

One of the most intriguing cars making the auto show circuit is India's REVA-NXG, built by the REVA Electric Car Company (RECC) in Bangalore. It is potentially the world's commuter/ shopper dream car. It's a beautiful all electric two-seat roadster with a range of 124 miles per charge and a top speed of 75mph. It uses a 50 horsepower AC induction motor to drive the front wheels and is powered by sodium-nickel-chloride (Zebra) batteries. It has ventilated disc brakes with regenerative braking. It is equipped with what the aviation industry calls a "glass cockpit"; RECC calls it a "wireless tablet." It's a touch-screen that displays all the vehicle's dashboard functions: speed, state-of-charge, range, GPS navigation, and MP3 music. It weighs 1715 lbs, is 105" long, 62" high, and 66" wide. It was designed by Dilip Chhabria Design Pvt Ltd to demonstrate India's technical capabilities. It would make the perfect commuting and errand car. It's as cute as a bug and should sell like hotcakes should REVA decide to market it.

A one-liter car is a car that uses one liter of fuel, about a quarter of a gallon, to propel the car 100 kilometers or 62 miles. In

2002, VW designed the 1-Liter Car, it weighed 290kg/638lb, and could go 100km on 0.89 liters of fuel. At the 2009 Frankfurt Auto Show, VW unveiled the follow-on L1, which weighs 308kg/838lb, and can go 100km on 0.99 liters of fuel. The carbon fiber body was developed in a wind tunnel and has a Cd of 0.159. Seating is tandem, but will probably be side by side in the production model. It has a two-cylinder, 39hp turbo-diesel engine and a 14hp electric motor. It has a top speed of 100mph, but at that speed, fuel economy drops to 1.38 liters per 100km, about a measly 170mpg. Its 6.5 liter tank gives it a range of about 650 kilometers. It is scheduled for production in 2010.

Toyota's concept 1/X (pronounced one-xth) has many interesting features, the body is constructed entirely of ultralight, ultrastrong carbon fiber reinforced plastic (CFRP). It has the same interior space as the Prius, but has a curb weight of 926 pounds compared to Prius's 2890 pounds. This mass decompounding enables the 1/X to be propelled by a 500cc engine versus the 1500cc engine required by the Prius. This, in turn, enables the 1/X to get 100mpg versus 46mpg for the Prius. The 1/X acceleration equals the Prius, and its small four gallon tank gives it a range of 600 miles. The 1/X is a PHEV, its battery pack uses Li-Ion batteries versus Prius's less efficient NiMH. The rear-wheel-drive power unit is located under the rear seat, where it takes up no interior or trunk space. The superstrong CFRP construction permits the A,B, and C-pillars to be much smaller, but stronger, greatly increasing all-around visibility and a feeling of openness. The roof is made of a bio-plastic material derived from kenaf and ramie plants. The material improves insulation, reduces noise, and increases the amount of light entering the cabin. The four ultralight, but comfortable, seats are knitted three-dimensionally from polyester fiber that functions as a spring and as a damper.

The concept 1/X is equipped with the same complicated and expensive series/parallel system as the Prius, which probably would be changed to the cheaper and simpler series system used by the Volt before going into production.

ENDGAME

The Middle-East is a Mare's Nest. Biofuels and PHEVs offer America an escape route out of this nightmare. During World War II, America launched the Manhattan Project to develop atomic weapons. The outcome was the end of the war with Japan and the birth of atomic energy. The cost was $20 billion in today's dollars. In the 1960s, America launched the Man to the Moon Project, which gave birth to space exploration and myriad advances in technology. The cost was $100 billion in today's dollars.

Set America Free (SAF) is proposing a Manhattan-type project to give America complete energy independence in four years at a cost of only $12 billion, which is probably optimistic. SAF is proposing an array of subsidies and tax incentives that would, for example encourage Detroit to develop and produce PHEV/FFV ultralight vehicles. Grant subsidies to the oil industry to install biofuel pumps in the existing fueling stations nationwide. Provide consumer tax incentives to buy PHEV/FFV vehicles. Mandate substantial incorporation of PHEV/FFV vehicles into federal, state, and municipal fleets. Provide tax incentives for corporate fleets and taxi fleets to switch to PHEV/FFVs. Allocate funds to build commercial scale demonstration plants to produce a variety of non-petroleum based fuels from a variety of sources, such as: cellulosic ethanols, butanol from biomass, methanol from coal, etc.

Energy storage devices such as batteries and ultracapacitors are the key technologies to the widespread adoption of PHEVs. All mass produced batteries are currently produced in Asia, with massive government support of research and production. America cannot afford to be dependent on foreign countries for the key components of its future transportation system. It would place us right back in the mess we're in today in the Middle-East. From the frying pan into the fire. We must set up long-term, multi-billion dollar government support for the development and production of technologies such as: batteries, ultracapacitors, fuel-efficient engines, high-performance electric motors, new materials, etc. It will cost America a few billion dollars a year, but it's chump change compared to what we have already spent in Iraq. It is difficult to see what America stands to gain in the long run from the Iraq/Afganistan mess. It is easy to see what all Americans can gain from a Manhattan-type energy program: energy independence and a new age of agricultural and industrial prosperity.

Power plants must produce electricity on demand. They have no way of storing electricity, so in a system called grid stabilization, utilities must pay billions of dollars a year to power plants (usually natural gas) to be on standby to produce the extra power needed to maintain system frequency at precisely 60 cycles per second. In a system called vehicle-to-grid (V2G), the batteries and supercapacitors in tens of millions of PHEVs and EVs could be utilized to store the extra electricity the utilities generate at night and off-peak hours. Most cars are parked about 20 hours a day. Utilities would use the parked cars to drawdown or store energy, eliminating the need to pay for standby power or build new power plants.

Google.org, the philanthropic arm of Google Inc, has launched RechargeIT (Recharge a Car, Recharge the Grid, Recharge the Planet). RechargeIT will issue over $10 million in RFPs to fund development, adoption, and commercialization of PHEVs, EVs, V2G, batteries, capacitors, materials, renewable energy, etc. Google has signed up an impressive group of collaborators including: The Brookings Institution, Electrical Power Research Institute (EPRI), Pacific Gas & Electric (PG&E), Enterprise Rent-A-Car, A123 Systems, and The Rocky Mountain Institute (RMI). PG&E is contracted to demonstrate the viability of V2G technology. Dr. Larry Brilliant, Executive Director of Google.org said V2G technology "can quadruple the fuel efficiency of cars on the road today and improve grid stability." EPRI is supporting PHEV and V2G R&D. RMI's Breakthrough Design Team worked with Alcoa, Johnson Controls-Saft (JCS), and the Turner Foundation to design a practical PHEV. RMI spinoff Bright Automotive figured the best way to extend vehicle range is to make them more aerodynamic and lighter.

Pursuant to this, Bright Automotive designed the IDEA, a lightweight, low-drag, commercial PHEV van. The Bright IDEA weighs only 3200 lbs (less than a Prius), has a Cd of 0.30, has a 30-mile all-electric range, and is able to deliver 100 mpg for the average customer. Typical current delivery vans get about 15 mpg. Bright has plans to produce 50,000 vehicles annually by 2013. PHEV vans and trucks can have a major impact on U.S. oil dependency. Federal, state, and local fleets own about four million vans or trucks; corporations own much larger fleets.

In the future Google predicts:

> OEMs will mass produce PHEVs and EVs.
> Biofuels will replace gasoline.

PHEVs and EVs will be parked in solar carports.
V2G systems will be operational nationwide; it is estimated each V2G PHEV might earn its owner as much as $3000 per year.

Eventually, the cars of the future will incorporate approximately the following technologies:

An ultralight/ultrastrong carbon composite and aluminum body. It will be a series hybrid powered by advanced electric motors, backed up by an ultracapacitor. It will employ a small, light, multi-fueled engine driving the generator to power the electric motors and recharge the lithium-ion battery pack and the ultracapacitor; augmented by regenerative braking. Mechanical systems will be replaced by drive-by-wire systems: brake-by-wire, steer-by-wire, suspension-by-wire, etc. All drive-by-wire systems will be redundant as on aircraft. All glazing will be made of new polycarbonates, similar to those currently used in headlight covers; which are 50% lighter than glass. To minimize unsprung mass, the two front electric motors and brakes will be mounted inboard, connected to the wheels by carbon fiber half shafts. The two rear electric motors will be light enough to be mounted within the wheel hubs. This produces a lower rear floor, better underbody aerodynamics, and eliminates the driveshaft, differential, and axle. It will be equipped with Michelin developed run-flat wheels and tires that can be driven at highway speeds on four flat tires.

The lightweight interior will be designed to meet safety, comfort, acoustic, thermal, and aesthetic requirements. The carbon fiber look has become de rigueur in high performance car interiors; thus, the inner surface of the carbon fiber safety cell will be left exposed. Another interior safety feature will be sidestick control of steering, braking, and acceleration. Sidesticks get rid of the steering column and pedals; the leading cause of injuries in collisions. This is especially true for women drivers, many of whom have to move the seat forward to reach the pedals. Sidestick control permits much faster reaction times and evasive maneuvers; all high performance aircraft utilize sidesticks. It will be equipped with another technology borrowed from the aircraft industry – a glass cockpit. The screen will display all dashboard functions and be controlled by voice commands. It will not be designed as a car with chips, but a computer with wheels. The auto industry of the future will be run from the Silicon Valleys of the world instead of the Detroits. Using the technologies outlined above, cars having approximately the following specifications can be built right now:

Passengers – 4
Curb Weight – 1500 lbs
Cd – 0.19
Range – 2000 miles (if equipped with a superbattery/ supercapacitor.)
Top speed – 150mph
0-60 time – 6 seconds
PHEV Mileage – 300mpg

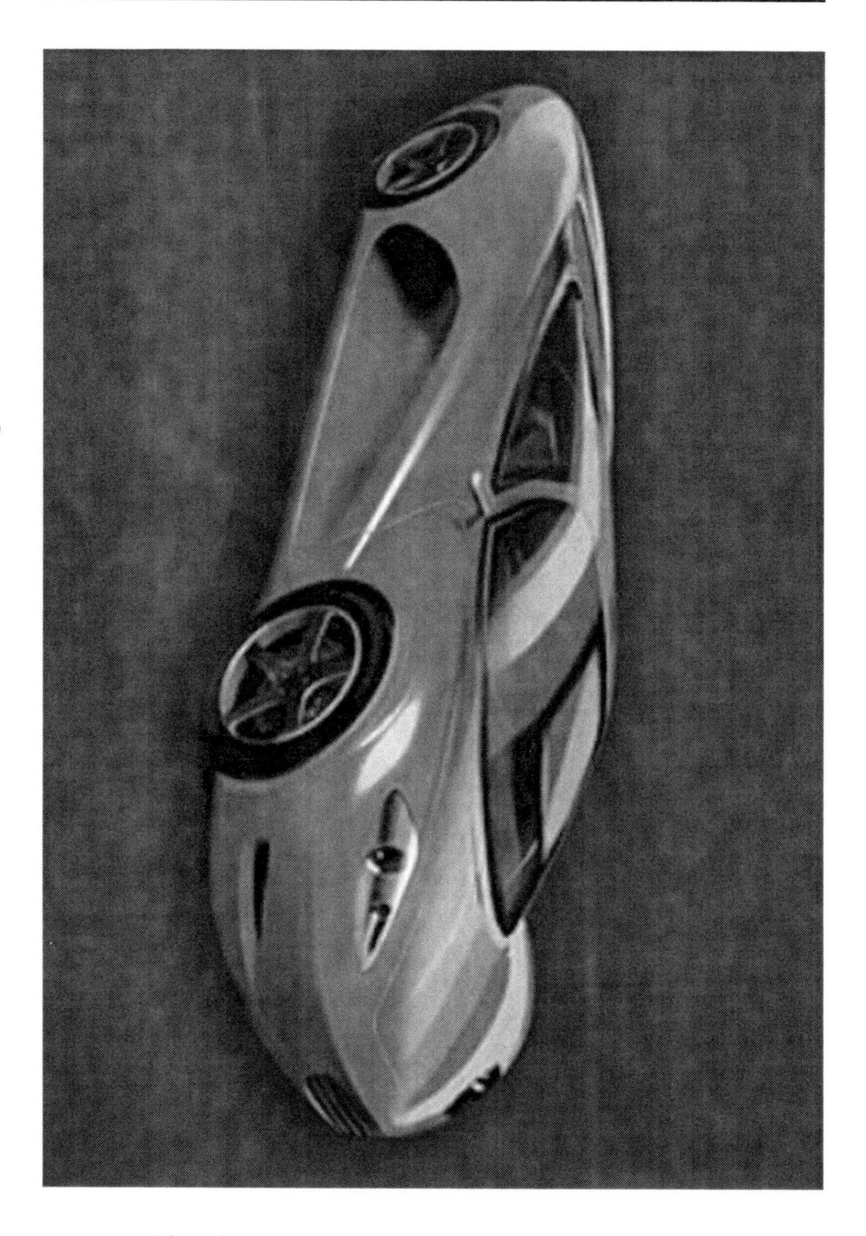

Velozzi Supercar Image courtesy Velozzi Group.

SOLO Image courtesy Velozzi Group.

APTERA Image courtesy Aptera Motors, Inc.

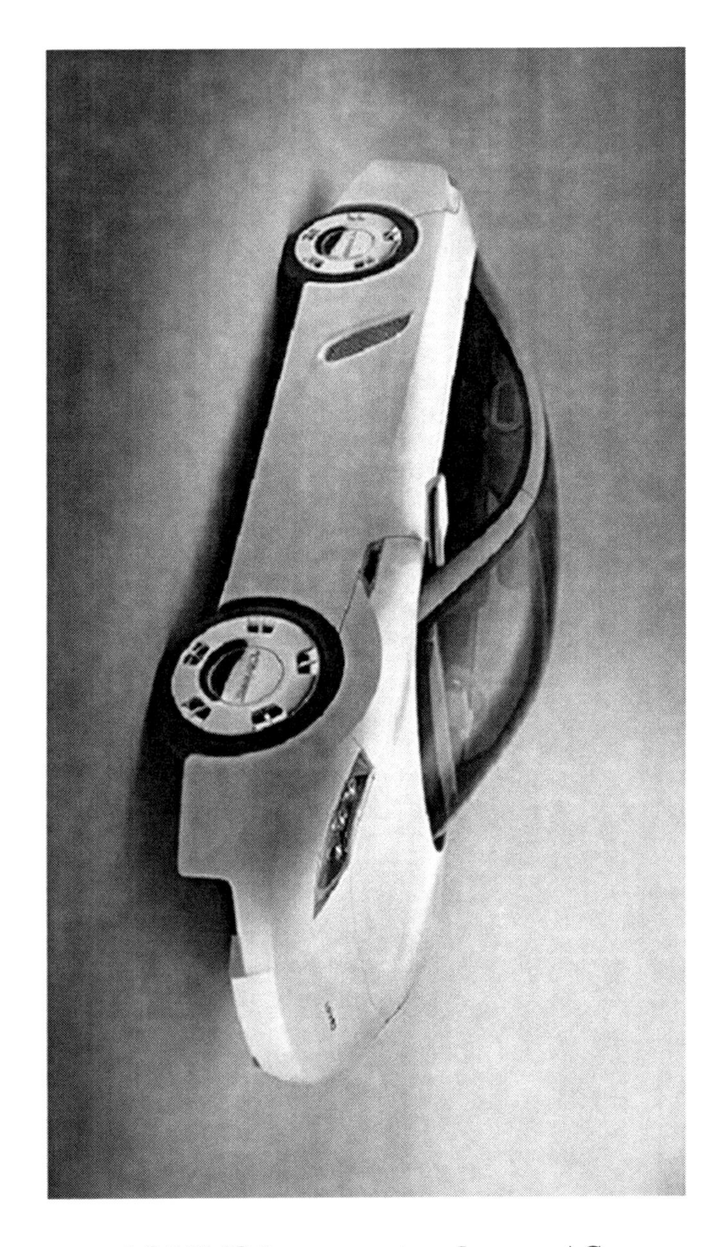

LOREMO Image courtesy Loremo AG.

FUEL CELLS

Reports of the death of fuel-cell cars have been exaggerated; Toyota and GM plan to put them in production by 2015. Hydrogen is our most abundant element. Hydrogen powered fuel celled vehicles would seem to be the answer to all our fuel problems. PEM (proton exchange membrane) fuel cells, the most common type, have a 40% fuel-to-power efficiency, compared to 20% for an internal combustion engine. Unlike batteries, fuel cells never lose their charge; as long as hydrogen and oxygen are available, they generate electricity. However, hydrogen storage is the greatest obstacle to getting fuel cell vehicles on the road. Being the lightest gas, hydrogen occupies too much volume, and it's flammable. For decades, the problem has been compact and safe storage.

The two most common methods of producing hydrogen are electrolysis of water and steam reformation of natural gas. Currently, electrolysis uses more energy than it creates; companies like GE are hard at work on this problem. Steam reformation of natural gas is already widely used in the glass, petroleum, and steel industries. The problem is that natural gas is relatively scarce and expensive. Aside from safe and compact storage, the challenges of building a national distribution infrastructure are enormous. Aside from the fact that the technology is not mature, there's the classic chicken-and-egg dilemma. Hydrogen producers are reluctant until the vehicles are on the road; automakers are reluctant until the distribution infrastructure is in place. The National Academy of Science estimates the U.S. would need to subsidize the hydrogen industry to the tune of $3 billion to $4 billion per year for 15 years to put the national infrastructure in place. A tidy sum, but its about the amount currently being spent subsidizing the corn-based ethanol industry.

The U.S. has about 170,000 gas stations. GM estimated that 11,700 on-site hydrogen natural gas reformers could put a station within 70% of the U.S. population, and permit refueling every 25 miles along the 130,000-mile National Highway System. Deutsche Shell said it could put hydrogen pumps in all its German stations in two years.

Notwithstanding the problems, research plows ahead all over the world. GE Global Research in Niskayuma, NY is operating an experimental low-cost electrolyzer to convert water to hydrogen and oxygen. Electrolysis is nothing new, water is mixed with an electrolyte, like alkaline potassium hydroxide, and made to flow past a stack of electrodes. Electricity causes the water molecules to split into oxygen and hydrogen gases. Currently, electrolyzers are hand assembled from expensive metals. GE is building their electrolyzers out of a GE plastic called Noryl that is easy to fabricate and is resistant to the electrolyte. To get more hydrogen out of a smaller electrode, GE borrowed a spray-coating they use fabricating GE jet engines parts. The electrodes are coated with a proprietary nickel-based catalyst that has a larger surface area. They plan to build a commercial version that will produce hydrogen for about $3/ kilogram.

China is positioning itself to become the world's largest producer of hydrogen; it is already number two. China is developing low cost in-situ extraction of methane and hydrogen from its vast coal resources, which will enable it to move energy by pipeline rather than by thousands of trains. Once hydrogen is introduced on a massive scale, the complete elimination of petroleum is inevitable; probably within 25-30 years. Royal Dutch/Shell, highly regarded worldwide for their long-range forecasting, predict China will leapfrog the world in mass production of superefficient cars and hydrogen fuel cells. Building superefficient carbon-fiber type cars is labor intensive; China has the world's largest supply of low-cost skilled labor and the labor pool is growing rapidly. China's huge pool of low-cost skilled labor will permit them to produce aerospace quality cars at Wal-Mart prices. Building PEM fuel cells is also labor intensive. RMI's Lovins and Williams have said the key to unlocking hydrogen's enormous potential lies in the early deployment of superefficient vehicles, which through mass decompounding, will lead to smaller and cheaper fuel cells and hydrogen fuel tanks. In all probability superbattery/supercapacitor development will preclude the need for fuel cells.

Some automotive futurists project that the industry first develop gasoline PHEVs, then biofuel PHEVs, then ultralight superefficient PHEVs, and finally fuel cell PHEVs. China may leapfrog this scenario. Hakko-Ichiu was the military/political program of Japan during World War II. It translated control (ichiu) of the world (hakko). Japan had a plan to control Asia and ultimately the world – the Greater East Asia Co-Prosperity Sphere. The plan supposedly had it roots in the Tanaka Memorial, presented to the Emperor by Premier/General/ Baron Gi-ichi Tanaka in 1927. Perhaps China has a new Tanaka Plan!

POSTSCRIPT

GOLDEN AGE of the AUTO

The 1930s were the "Golden Age" of car design and biplane design. It was a magic period, during which the most beautiful automobiles and military biplanes ever designed were produced. In the 1930s, if you were really rich and wanted a really elegant and unique car, you went to somebody like Bugatti, Duesenberg, Delage, Alfa Romeo, or Mercedes. They sold you a complete chassis and running gear (engine, transmission, wheels, etc.); everything except the body. You then had your "Duesy" or Delage chassis delivered to a renowned coachbuilder like Walker-LeGrande or Letourneur et Marchand. The coachbuilder would design and build an incredibly gorgeous and unique body and mount it on your Duesenberg or Delage chassis. The result would be a work of art like the 1932 Duesenberg SJ by Walker-LeGrande or the 1938 D8 120 Delahaye/Delage by Letourneur et Marchand.

For some reason 1937-38 were the "Golden Years" of the "Golden Age" producing such "things of beauty" as:

1937 Alfa Romeo 2900 B, body by Carrozzeria Touring
1938 Talbot-Lago, "guotte d'eau", body by Figoni et Falaschi
1937 Mercedes 540 K, body by Sindelfingen

All were such magnificent works of art that they should have been delivered to the nearest gallery of art; but we only know that now. World War II ended all that, just as World War I ended civilization.
The golden years of the 1930s were a rich man's game; the 1960s and 70s gave the poor guys a shot. Volkswagen was building the modern equivalent of Henry Ford's Model T – the Bug. This was a cute, cheap, simple, reliable little car that the whole world was buying by the millions. Connoisseurs of cars aka car nuts, particularly in Southern California, noticed that all you had to do was unscrew 14 bolts and a few wires, lift off the Bug body, and voila/ecco/behold, you had a neat, compact, complete chassis

and running gear. The poor man's "Deusy", ready for a wild new body. Poor guys started setting up shops to build exotic bodies out of fiberglass. Designs ran the gamut from dune buggies to exotic originals to copies of Italian carrozzeries like Ital Design. These little backyard carrozzeries became the "kit car" industry; it's still going strong.

If the auto industry goes the route of GM's Skateboard, there will be a great renaissance of the coachbuilder/carrozzeria industry. The Skateboard, like the VW Bug, will provide a complete standard chassis that you can bolt your body to. Of course, GM, Ford, Toyota, et.al., will provide their own line of bodies like sports car, sedan, truck, etc. for their chassis, but these skateboards will open a new world for the independent coachbuilder. These independents will be able to offer customers a galaxy of exotic bodies to mount on the skateboards. Two fabulous examples come immediately to mind. David Clash in Australia builds the Devaux. The Devaux beautifully incorporates all the best design features of the:

1937 Alfa Romeo 2900 B body by Carrozzeria Touring
1938 Talbot-Lago teardrop body by Figoni et Falaschi
1937 Delahaye 135 M Cabriolet body by Guilloie

Or if you prefer a more beautiful than the original 1938 Talbot-Lago by Figoni et Falaschi, George Balaschak of Palm Beach Gardens, Florida is your man.

GLOBAL WARMING or PERIODIC DOOMSDAY CRUSADE?

We are in the midst of a periodic doomsday crisis; the latest being the greenhouse gas/global warming crusade. Today's "apocalypse now" "Mahdi" is Al Gore; in the 1960s, it was Paul Ehrlich, who issued such doomsday pronouncements as:

> "The battle to feed humanity is over. In the 1970s and 1980s hundreds of millions of people will starve to death in spite of any crash programs embarked upon now."
> Paul Ehrlich, *The Population Bomb,* 1968.
> "Hundreds of millions of people will soon perish in smog disasters in New York and Los Angeles --- the oceans will die of DDT poisoning by 1979 --- the U.S. life expectancy will drop to 42 years by 1980 due to cancer epidemics."
> Paul Ehrlich, *Ramparts*, 1969.
> "--- nutritional disaster seems likely to overtake humanity in the 1970s (or, at the latest, the 1980s) --- A situation has been created that could lead to a billion or more people starving to death."
> Paul Ehrlich, *The End of Affluence*, 1974.

Whether or not Al Gore is another Paul Ehrlich only time will tell. I subscribe to Michael Crichton's view of the latest "crisis":

> "Has it ever occurred to you how astonishing the culture of Western society really is? Industrialized nations provide their citizens with unprecedented safety, health, and comfort. Average life spans increased 50 percent in the last century. Yet modern people live in abject fear. They are afraid of strangers, disease, of crime, of the environment. They are afraid of the homes they live in, the food they eat, the technology that surrounds them. They are in particular panic over the things they can't even see – germs, chemicals, additives, pollutants. They are timid, nervous, fretful, and depressed. And even more amazingly, they are convinced that the environment of the entire planet is being destroyed

> around them. Remarkable! Like the belief in witchcraft, it's an extraordinary delusion – a global fantasy worthy of the Middle Ages. Everything is going to hell, and we must all live in fear."
>
> Michael Crichton, *State of Fear*, 2004.

For the past 800,000 years, there have been periods of about 100,000 years called Ice Ages. These are followed by periods of about 10,000 years called Interglacial Periods; followed by another Ice Age. We're about 10,500 years into the current Interglacial Period; meaning we're about 500 years overdue for another Ice Age. If CO_2 does in fact cause global warming, logically everyone should go out and buy a Hummer. From AD 800 to 1300, we had the Medieval Warm Period, when Greenland was actually green. This was followed by the Little Ice Age from AD 1350 to 1900. The Medieval Warm Period was a time of peace and plenty; the Little Ice Age of war and starvation. In the 1970s, environmentalists were hysterical about an impending New Ice Age.

Environmentalists proclaimed the 1990s to be the "hottest decade", producing the infamous "Hockey Stick" graph to show an exponential rise in global temperatures during that period. What they failed to mention was that the Soviet Union collapsed and closed hundreds of weather stations across Siberia. When you shut down the measuring stations in one of the coldest areas of world, covering ten time zones, the average measured global temperature can go no where but up. It is interesting to note that the temperatures did not rise in the Southern hemisphere of the planet during this period.

According to Dr. Richard S. Lindzen of MIT, reputed to be "the most renowned climatologists in all the world" all the greenhouse gases together, including CO_2 and methane, produce less than two percent of the greenhouse effect. The lower atmosphere is composed of nitrogen (78%), oxygen (21%), water vapor (4%), and CO_2 (0.0368%), i.e. 368 ten thousandths of 1%. About 10% of global CO_2 emissions come from cars, SUVs, and pickup trucks, that is, 10% of 0.0368%! Water vapor is the dominant greenhouse gas, accounting for 95% of the Earth's greenhouse effect. It is 99.999% natural and thus, "beyond the reach of man's screwdriver." The 1997 Kyoto conference found that it could not reduce water vapor, thus, it could not reduce global warming. But being good government bureaucrats, they could not vote to do nothing. In *Cool It,* economist Bjorn Lomborg points out that cold weather kills far more people than hot weather. In Europe, about 200,000 die each year from excess heat, whereas, 1.5 million die each year from excess cold. In Britain, it is estimated that a 3.6°F increase will mean 2000 more heat deaths, but 20,000 fewer cold deaths. Lomborg says "global warming may cause a decrease in mortality rates, especially of cardiovascular diseases."

Since the peak of the last ice age about 18,000 years ago, sea levels have risen, on average, about seven inches a century; there was no acceleration during the 20th century. Ocean warming increases evaporation and precipitation, raising global supplies of fresh water. Agronomists agree that more CO_2, an essential plant food, would enhance the growth of crops and forests, and perhaps delay or cancel the next ice age. Many scientists believe sunspot activity and the earth wobbling on its axis are the main causes of climate change. Sunspot activity has reached a 1000-year high; more sunspots mean warmer weather; fewer mean colder. There is global warming on Mars; since there are no people on Mars, it must be due to the sun. About 655 million years ago the Earth experienced the most rapid global climate change in recorded history called the "Paleocene-Eocene Thermal Maximum." The ocean temperature was 18 to 27 degrees hotter than it is today. Antarctica was a balmy temperate zone covered in beech trees and ferns. Just 400,000 years ago, Greenland was covered in forests and basking in temperatures an estimated 27 degrees F warmer than today. Which raises the question: what is the "correct" Earth temperature?

DDT was developed in World War II to combat insect-borne diseases like typhus and malaria. By the 1960s, malaria had been almost totally eradicated from the planet. In 1962, Rachel Carson wrote her pseudoscientific book *Silent Spring*, calling for the worldwide ban of DDT. Organizations like the Audubon Society and the Sierra Club succeeded in getting the World Health Organization (WHO) to ban DDT, thus depriving billions of poor, helpless people of their only practical weapon against malaria. As a result, for the next 30-odd years, approximately three million people a year died from malaria worldwide. This made DDT's banners responsible for over 90 million preventable deaths, putting them on a par with Stalin and Mao.

Trying to build climate models that could incorporate such myriad and massive factors as sunspot activity, earth wobbling on its axis, changes in global air and ocean currents, volcanic activity, tectonic plate movement, etc. are beyond the capabilities of all of the algorithms and supercomputers we are ever likely to design or build. We know from chaos theory that even if you had a perfect model of the world, you would need infinite precision to predict future climate.

As the authors of *Unstoppable Global Warming: Every 1,500 Years* said: "We are being humbugged by activists of whom we should automatically be wary. We are also being humbugged by the journalists we pay to provide us with reliable information, healthy skepticism, and differing perspectives. Now we are being humbugged by highly trained professional scientists, many of them working on government research grants."

Despite the Supreme Court's decision, CO_2 is not a pollutant. Aside from oxygen, on which every living creature depends, CO_2 is our second most important gas. Due to its ability to harness the sun's energy through photosynthesis, it maintains every form of vegetation on Earth. Has anybody stopped to think what would happen if we seriously reduced CO_2 emission on the planet? Not that we can do it, but suppose we did. All plant life lives on CO_2, *that's what they breathe.* We would be suffocating all the food crops, not to mention the trees, plants, and flowers. Aside from mass starvation, the whole planet would become a barren dust bowl like the moon. Besides, suppose it does warm up a bit. What's wrong with Canada and Siberia being like Brazil? It would do them both a world of good. Brazil has a growing surplus of ethanol. The U.S. imposes a 54 cents per gallon tariff on Brazilian ethanol and zero cents per gallon on Middle Eastern petroleum. We should stop discriminating against poor Brazilian farmers while we subsidize Saudi billionaires and fire-breathing Iranian mullahs. Brazil is a beautiful country, filled with fabulous flowering trees, covered in orange or purple blossoms, like the Espatodia and the Quaresmeira. The trees and skies are filled with fantastic birds, dressed from beak to tail in all the colors of the rainbow. And the beaches are filled with fun-loving beautiful people. Canada and Siberia should be so lucky. With Canada and Siberia transformed into tropical countries, the world would be inundated in an abundance of food. There would be no hunger. The globe would be awash in biofuel. Tropical countries, like Brazil, get two crops a year. Petroleum, if needed at all, would only be used for petrochemicals. The Muslims could put a fence around the Middle-East and retreat to the seventh century. So let's not be in a hurry to suffocate the trees!

Speaking of saving the trees, left-wing enviro-authoritarians have been recently inspired by nine-time Grammy winner and environmental braintruster Sheryl Crow's brilliant plan to save the world's forests. The brilliance of the plan lies in its simplicity. To quote spokesperson Crow: "Although my ideas are in the earliest stages of development, they are, in my mind, worth investigating. One of my favorites is in the area of forest conservation which we heavily rely on for oxygen. I propose a limitation be put on how many squares of toilet paper can be used in any one sitting. Now, I don't want to rob any law-abiding American of his or her God-given rights, but I think we are an industrious enough people that we can make it work with only one square per restroom visit, except, of course, on those pesky occasions where 2 or 3 could be required."

If you look at the globe from the top, you quickly realize that the countries you think of as being half a world apart, are in reality, amazingly close neighbors. The thawing of the Arctic Ocean will create the fabled Arctic Bridge. Instead of being crowned by a mass of floating ice, in summer, the planet will have a blue top, and an open sea five times the size of the Mediterranean. A Northern Sea Route will shorten the maritime journey from Northeast Asia to Europe by 40 percent. The 450-year search for the Northwest Passage, the shortcut from Europe to Asia across the top of Canada, will be attained. Churchill and Murmansk will become sister cities. Shipments from Murmansk to Ontario, now go through the St. Lawrence Seaway and the Great Lakes, and take 17 days. The voyage from Murmansk to Churchill takes eight days, with rail links from Churchill through the American Midwest all the way to Monterrey, Mexico. Aside from the global bonanza in shipping costs and times, the U.S. Geological Survey estimates that 25% of the world's undiscovered oil and natural gas are in the Arctic. The natural resources of the Arctic are calculated to be in the trillions of dollars.

CAP-AND-TRADE

Guess what! The whole cap-and-trade flim-flam was first cooked-up by those wonderful guys at Enron. They saw it as a golden opportunity to make incredible fortunes running the emission permits market. The plethora of exotic derivative instruments would metastasize exponentially. In 2007, AIG CEO Martin Sullivan said: AIG "can help shape a broad-based cap-and-trade legislative proposal, bringing to this critical endeavor a unique business perspective on the business opportunities and risks that climate change poses for our industry." Translation: This will give us a license to steal. Goldman Sachs was AIGs single largest partner for credit default swaps, and the biggest corporate cheerleader for cap-and-trade. Credit default swaps were based on the inherent value of houses. Home prices may crash by 50% or more, but they will never go to zero. Emission permits have an intrinsic value of zero, which is exactly what the will be worth when the inevitable cap-and-trade crash comes.

The American Clean Energy and Security Act (ACES), a.k.a. cap-and-trade, seeks to reduce projected CO_2 levels by 50% between 2010 and 2099. According to Iain Murray of the Competitive Enterprise Institute (CEI), this would require cutting CO_2 production by 41.1 billion tons for each of the 90 years between 2010 and 2100. According to CEI, this could theoretically be done by pursuing any one of the following draconian options:

Plant trees on barren land covering about 14 million square miles; an area four times the size of the U.S.
Build an additional 5589 one-gigawatt atomic power plants.
Build 11,220 "zero-emission" 500-megawatt coal-fired power plants.
Increase current wind turbine capacity 575 times.

President Obama told the San Francisco Chronicle cap-and-trade is designed to "bankrupt" coal-fired power plants. Chip Knappenberger of the Institute of Energy Research has estimated ACES would reduce global warming by nine-hundreds of one degree Fahrenheit in 90 years. President Obama says the price of electricity will "necessarily skyrocket" when cap-and-trade is implemented. It is estimated the cost of electricity will increase 77% to 129%; gasoline will go up 60% to 144%; home heating oil and natural gas will go up 100%. The National Association of Manufacturers (NAM) estimates cap-and-trade will cost three million to four million jobs; the Heritage Foundation estimates 1.8 million to 5.3 million; other estimates run to seven million. Lawrence Kudlow calls cap-and-trade "a malthusian plan for root-canal austerity." That's the good news. The bad news is cap-and-trade will bring on global economic collapse on a scale that will make the recent Dot.Com, McMansion, Hedge Fund and Derivatives disasters look like a Sunday Ice Cream Social. The best way to rescue the economy would be to halt the Federal government's relentless march toward implementation of the Bureau of Economic Planning and National Resources Directive # 10-289, as outlined in Ayn Rand's *Atlas Shrugged.*

It will be Ringling Brothers-Barnum & Bailey on a global scale. Chicanery, skullduggery, flim-flam, fraud, and deception will rule supreme. Tinhorn Napoleon government bureaucrats, environmental crazies, robber-baron industrialists, and a new breed of "masters of the universe" traders will take over the planet. Imagine an international army of matastasized hedge fund and derivative royalty. All the Charlie Ponzis, Ivar Kruegers, and Bernie Madoffs will be in charge of the world economy. As always Charles Krauthammer put it perfectly: "The last time the selling of pardons was prevalent - in a predecessor religion to environmentalism called Christianity - Martin Luther lost his temper and launched the Reformation." This time there will be total collapse of the world economic system and global chaos.

Global Warming is becoming a third rail like Social Security. From the above it is not difficult to figure out my opinion on the subject. But it does not make a damn which side is right. Whether you're a "doubting Thomas" or a wild-eyed "tree-hugger," the petroleum problem must be resolved and quickly. Right now, most of the world's petroleum is in the hands of crazy people who are using the oil money to fund World War IV. And it won't get better or go away as long as the world's transportation systems run on their product. By pursuing the technologies outlined in the Introduction to this book, world dependency on petroleum can be completely eliminated and CO_2 emissions enormously reduced. It would cut off the funding of religious fanatics bent on ruling the world. Why anyone would want to or think that they could is one of history's more interesting questions.

To paraphrase RMI's Amory Lovins:

Imagine a revitalized and globally competitive American automobile industry producing a new generation of revolutionary highly efficient, safe, durable, fun-to-drive vehicles that customers don't have to be bribed to buy. Imagine the sweeping economic benefits of a new materials industry mass producing cheap, ultrastrong, ultralight products like cars, trucks, trains, aircraft, ships, building materials, etc. Imagine a secure national fuels infrastructure based entirely on U.S. energy resources and on dynamic rural communities producing biofuels, plastics, windpower, solarpower, etc. Envision completely new industries and millions of high paying jobs being created while the use of petroleum plummets toward zero. Imagine our military concentrating on its core mission of defending America instead of protecting oil supplies in the hostile Middle East. Imagine Americans enjoying the largest and broadest-based tax cuts in history. Envision America regaining the moral high ground internationally.

Sound utopian? It is not.

REFERENCES

A ravenous dragon, The Economist, 3/15/08.

Senator Lamar Alexander, *The real reason for fear,* The Washington Times, 10/9/09.

As Augustine Sees It, Aviation Week & Space Technology, 5/11/09.

Arnaud de Borchgrave, *Confusion reigns supreme,* The Washington Times, 6/17/09.
Arnaud de Borchgrave, *Finance mavens gloomy,* The Washington Times, 11/27/09.

Kevin Bullis, *A Preassembled Nuclear Reactor,* Technology Review, 6/16/09.

Accelerated Composites, *A 330 MPG Car for Everyone,* 1/18/06.

Batteries now included, The Economist, 3/14/09.

Alliance Bernstein, *The Emergence of Hybrid Vehicles, Ending Oil's Stranglehold on Transportation and the Economy,* Research on Strategic Change, June 2006.

Beyond Ethanol, Popular Science, February 2008.

Biofuels Bonanza: Exxon, Venter to Team up on Algae, The Wall Street Journal, 7/14/09.

Giovanni Bisignani, *We Are Misunderstood and It's Our Own Fault,* Aviation Week & Space Technology, 4/16/07.

Sherry Boschert, *Plug-in Hybrids, The Cars That Will Recharge America,* New Socirty Publishers, Canada, 2006.

Christine Blackman, *Scientists discover 9-second lithium-ion recharge,* Cleantech, 3/13/09.

Joel K. Bourne, Jr., *Green Dreams,* National Geographic, October 2007.

Michael Briggs, *Widescale Biodiesel Production from Algae,* University of New Hampshire, August 2004.

Michael M. Brylawski and Amory B. Lovins, *Ultralite-Hybrid Vehicle Design: Overcoming the Barriers to Using Advanced Composites in the Automotive Industry.*
Rocky Mountain Institute, Snowmass, CO.

Brilliant 10, Popular Science, November 2009.

Kevin Bullis, *Clean Diesel from Coal,* Technology Review, 4/19/06.

H. Sterling Burnett, *Coal power boom,* The Washington Times, 12/26/06.

Biofuels from Switchgrass: Greener Energy Pastures. Oak Ridge National Laboratory, Oak Ridge, TN. eere@ornl.gov. bioenergy.ornl.gov.

John Carey, *Ethanol Is Not the Only Green in Town,* Business Week, 4/30/07.

Alan Caruba, *Supremely bad decision*, The Washington Times, 4/6/07.

Center for American Progress:

Resources for Global Growth, 2005.
Energizing Rural America, 2007.
Fueling a New Farm Economy, 2007.

Nick Chambers, *New Battery Alternative Stores Huge Amounts of Energy,* Batteries, Science, Ultracapacitors, 9/26/08.

Mark Clayton, *"Toyota moves to corner the 'plug-in' market."* The Christian Science Monitor, July20, 2006.

Michael V. Copeland, *The Hydrogen Car Fights Back,* Fortune, 10/26/09.

David R. Cramer, David F. Taggart, Hypercar, Inc., *Design and Manufacture of an Affordable Advanced-Composite Automotive Body Structure,* 2002.

Csaba Csere, The Steering Column, Car and Driver, March 2007.

Michael Crichton, *State of Fear*, Harper Collins, NYC, 2004.

Tom Clynes, "The Energy Fix, 10 Steps to End America's Fossil-Fuel Addiction." Popular Science, July 2006.

Danish Researchers Reveal New Hydrogen Storage Technology, Science Daily, 5/30/06.

DOE, EIA, *Crude Oil and Total Petroleum Imports Top 15 Countries,* 4/17/07.

Roger Doyle, *Food Boom,* Scientific American, May 2007.

K. Eric Drexler, *Engines of Creation,* Anchor Press/Doubleday, NYC,1986.

Paul Driessen, *Global warming insanity,* The Washington Times, 9/12/07.

Gail Edmondson, *The Race To Build Really Cheap Cars,* Business Week, 4/23/07.

Pete Engardio, *Can The Future Be Built In America?,*Business Week, 9/21/09.

Elaine Engeler and Alexander G. Higgins, *Energy answers sought in Earth's crust,* The Washington Times, 8/13/07.

Katie Fehrenbacher, *15 Algae Startups Bringing Pond Scum to Fuel Tanks,*

David A. Fulghum, *Striking Advantage,* Aviation Week & Space Technology, 6/4/07.

McKenzie Funk, *China's Green Evolution,* Popular Science, August 2007.

The Future of Energy, Popular Science, July 2009.

Robert Greenstein, *Poverty rate rises, health care dips,* The Washington Times, 9/15/09.

Andrew S. Grove, *What Detroit Can Learn From Silicon Valley,* The Wall Street Journal, 7/13/09.

Andy Grove in Washington Post; Volt, Tesla Other Reports, Cal Cars, 7/13/08.

Andy Grove Calls for Concentrated US Effort to Convert Pickups, SUVs and Vans to 40+ Mile PHEVs, Green Car Congress, 7/22/08.

Michael Grunwald, *The Clean Energy Scam,* Time, 4/7/08.

Green Gasoline Could Power Future Cars and Jets, Renewable Energy World, 4/10/08.

Tyler Hamilton, *Battery Breakthrough?* Technology Review, 1/22/07.

Paul Hawken, L. Hunter Lovins, *Natural Capitalism: Creating the Next Industrial Revolution*, Rocky Mountain Institute, 1999.

Richard Heinberg's Museletter: *The Food and Farming Transition,* 11/1/08.
" " " : *Energy Limits to Growth - Part III: Integrating Energy Sources,*3/20/09.

Ben Hewitt, *The 110-Volt Solution*, Popular Mechanics, May 2007.

Mike Hogan, *Time for Natural-Gas Autos?,* Barron's, 8/31/09.

Ian Hore-Lacy, *Small nuclear power reactors,* Encyclopedia of Earth, 4/5/09.

Christopher C. Horner, *The Politically Incorrect Guide to Global Warming and Environmentalism,* Regnery Publishing, Inc., Washington, DC, 2007.

Sam Jaffe, "Independence Way." Washington Monthly, July/ August 2004.

George Johnson, *Plugging Into The Sun,* National Geographic, September 2009.

Vinod Khosla, *Imagining the Future of Gasoline: Separating Reality from Blue-sky Dreaming?* September 2006.
Vinod Khosla, *A Near Term Energy Solution.* September 2006.
Vinod Khosla, *Biofuels Trajectory to Success.*

Vic Kolenc, *Company cultivates algae to make fuel,* El Paso Times, 9/27/07.

Felix Kramer, California Cars Initiative, *The 21st Century Car Industry: Fixed in the USA,* 8/11/09.

Clifford Krauss, Steven Lee Myers, Andrew C. Revkin, Simon Romero, *Arctic riches coming out of the cold*, The New York Times, 10/10/05.

Charles Krauthammer, *Limousine Liberal Hypocrisy,* Time, 4/26/07.

Emily Lambert, *Flivver Beats Hummer,* Forbes, 8/13/07.

Marianne Lavelle, *"The New Oil Rush."* U.S. News & World Report, 4/24/06.

Sean Lengell, *Treasury chief urges regulation of derivatives,* The Washington Times, 12/3/09.

Paul Johnson, *Where Industry Has Failed Us,* Forbes, 2/11/08.

Gary Kendall, *Plugged In: The End of the Oil Age,* WWFN, 2008.

Vic Kolenc, *Company cultivates algae to make fuel,* El Paso Times, 9/27/07.

Fred Krupp, *The Mother Lode,* Fast Company, April 2008.

Lawrence Kudlow, *Why China beats the U.S..* The Washington Times, 7/14/09.

Investing in Algae Biofuel, Green Chip Stocks, Angel Publishing, March 2008.

Martin LaMonica, *Joule adds CO_2 to sunlight to make fuel,* Green Tech, 7/26/09.

Marianne Lavelle, *The New Oil Rush,* U. S. News & World Report, 4/24/06.

Jay Leno, *The 100-Year-Old New Idea*, Popular Mechanics, May 2007.

Long debate ended over cause, demise of ice ages--may also help predict future, Oregon State University, 8/6/09.

Amory B. Lovins, *Hypercars: The Next Industrial Revolution.* The Hypercar Center, Rocky Mountain Institute, Snowmass, CO. hypercar@rmi.org.

Amory B. Lovins, E. Kyle Datta, Odd-Even Bustnes, Jonathan G. Koomey, and Nathan J. Glasgow, *Winning the Oil Endgame*, Rocky Mountain Institute, Snowmass, CO, 2005.

Jenny Mandel, *Biofuels: Algae producers bloom in search for Oil Alternative,* Greenwire, 10/27/08.

Frank Markus, *booze clues,* Motor Trend, April 2008.
Frank Markus, *add venture,* Motor Trend, January 2010.

Richard Martin, *The New Nuke,* Wired, January 2010.

Clifford May, *Paving the road to energy security*, The Washington Times, 1/26/07.

Michael Mecham, *Three- Step Process*, Aviation Week & Space Technology, 5/28/07.

James B. Meigs, *The Ethanol Fallacy*, Popular Mechanics, February 2008.

Hugo Miller, *Arctic schemes,* The Washington Times, 2/28/08.

Steve Milloy, "*Day of reckoning for DDT?,* The Washington Times, 9/25/06.

Robert Milliken, *Australia's coal question, The World in 2008,* The Economist.

Frank Morring, Jr, *Space Solar Power,* Aviation Week & Space Technology, 8/20/27/07.

New carbon material shows promise of storing large quantities of renewable electrical energy, EurekAlert!, University of Texas at Austin, 9/16/08.

Andrew Nikiforuk, *Electric cars: The wheel deal?,* Canadian Business, 3/30/09.

James Ott, *Algae Advances,* Aviation Week & Space Technology, 3/17/08.
James Ott, *Capturing* CO_2, Aviation Week & Space Technology, 3/17/08.

Randall Parker, *Chevy Volt Battery Overengineered Due to Unknowns,* Future Pundit, 4/13/09.

Kevin Phillips, *Bad Money,* Viking, NYC, 2008.

Bill Powell, *China's Amazing New Bullet Train,* Fortune, 8/17/09.

Wesley Pruden, *Emissions Control, we have a problem,* The Washington Times, 4/24/07.

Richard W. Rahn, *More tax oppression,* The Washington Times, 7/2/09.

Renewable Energy, The Washington Times, 3/4/08.

Rebecca Renner, *Green Gold in a Shrub*, Scientific American, June 2007.

David Sandalow, *Ending Oil Dependence,* 1/22/07.

Amanda Schaffer, *Breeding the Oil Bug,* Popular Science, April 2008.

George P. Schultz and R. James Woolsey, *Oil and Security*, The Committee on the Present Danger, Washington, DC, 2005.

Don Sherman, *Volt of inspiration or flash in the pan?* Automobile, April 2007.

Dennis Simanaitis, *Fueling Our Mobility*, Road & Track, November 2006.

Climate Change, CO_2 and the Automobile, Road & Track, June 2007.

S. Fred Singer and Dennis T. Avery, *Unstoppable Global Warming: Every 1,500 Years,* reviewed by Larry Thornberry, The Washington Times, 1/14/07.

Gunjan Sinha, *Pumping Coal,* Scientific American, May 2006.

Fred L. Smith Jr and William Yeatman, *Cap and traitors,* The Washington Times, 7/5/09.

Guy Sorman, *Economics Does Not Lie,* City Journal, Summer 2008.

Christopher Steiner, *Son of Wankel,* Forbes, 4/23/07.

Ben Stewart, *"Why not one hundred MPG?"*, Popular Mechanics, August 2006.

David Talbot, *Hydrogen on the Cheap*, Technology Review, May/June 2006.

The Green 50, Inc. Magazine, November 2006.

Justin Thomas, *Electric Mini: 0-60 in 4 Seconds: It Has Motors In Its Wheels,* 8/30/06, treehugger.com.

John J. Tkacik, *China: No longer “”””” rising* ,The Washington Times, 10/8/09.

Pat Toensmeier, *“Size Matters.”* Defense Technology International, May/June 2006.

William Tucker, *The Case for Terrestrial (a.k.a. Nuclear) Energy,* Imprimis, February 2008.

R. Emmett Tyrrell Jr, *Freedom from foreign oil,* The Washington Times, 10/2/09.

US Borrows a Billion Dollars Every Working Day to Import Oil, Fortune, 9/3/07.

U.S. Senate Committee On Finance, Subcommittee On Energy, Natural Resources, And Infrastructure Concerning Advanced Technology Vehicles: The Road Ahead, May 1, 2007.

Phillip Baxley, President, Shell Hydrogen LLC
Martin Eberhard, CEO, Tesla Motors, Inc.
Dr. Walter McManus, University of Michigan
David Vieau, CEO, A123Systems

Tim Vintura, *The Massive Yet Tiny Engine, 5/12/06.*

John Voelcker, *Lithium Batteries for Hybrid Cars*, IEEE Spectrum.

Vivek Wadhwa, *Engineering Gap? Fact and Fiction,* Business Week, 7/10/06.

Why Global Warming May Save Lives, Discover, 9/07.

Tom Whipple, *The Peak Oil Crisis: A Disruptive Technology,* 8/12/09.
" " , " " " " : *More Disruptive Technology,* 8/20/09.

George Will, *The Biofuel Follies,* Newsweek, 2/11/08.

Walter E. Williams, *Things to think about.* The Washington Times, 5/17/07.
Walter E. Williams, *Black Education,* Townhall.com, 12/22/09.

Alastair Wood, *Stimulus and science,* The Washington Times, 2/15/09.

R. James Woolsey, Testimony to U.S. House of Representatives Select Committee on Energy Independence and Global Warming Hearings on Geopolitical Implications of Rising Oil Dependence and Global Warming on 4/18/07 as reported by Felix Kramer at fkramer@calcars.org.

J. Patrick Wright, *On A Clear Day You Can See General Motors,* Avon Books, NYC, 1979.

Mark Zepezauer and Arthur Neiman, *Oil and Gas Tax Breaks: $2.4 billion a year, from the book Take the Rich Off Welfare,* Odonian Press, 1996.

Mortimer B. Zuckerman, *Getting Serious About Oil*, U.S. News & World Report, 8/7/06.
Mortimer B. Zuckerman, *The Energy Emergency,* U.S. News & World Report, 9/10/07.

Lisa Zyga, *How a Solar-Hydrogen Economy Could Supply the World's Energy Needs,* 8/24/09.

.

BOOK SUMMARY

Energy is the biggest business in the world; electricity is the energy of the future. Nikola Tesla's dream finally will be realized; the world is on the threshold of a new age of electricity. There is an arsenal of new technology magic bullets primed to give America total energy independence and economic revitalization. Reindustrialization and massive electrification will launch a new era of economic prosperity. America imports 70% of its oil at a cost of $700 billion per year - the greatest transfer of wealth in the history of the world. America can achieve total energy independence and economic rebirth by pursuing the following technologies: PHEVs, superbatteries/supercapacitors, algae bio-fuel, uncapped old oil wells, switchgrass, oil shale, natural gas, nuclear, solar, wind, geothermal, green gasoline, etc.

In the past 150 years (since 1859), the entire world has consumed one trillion barrels of oil.
America consumes 140 billion gallons of gasoline per year.
America has two trillion barrels of residual oil in 400,000 capped-off oil wells.
Beneath the Rocky Mountains, in the vast Bakken Formation, America has two trillion barrels of light, sweet crude oil.
Beneath the Appalachian Mountains, America has 2000 trillion cubic feet of natural gas; enough to last 100 years.
The Canadian tar sands contain over two trillion barrels of oil.
America has 275 billion tons of coal, enough to last 300 years; convertible to trillions of barrels of diesel oil.
There are over 1700 1000 megawatt coal-fired power plants in the U.S. that could produce 170 billion gallons of bio-fuel per year if equipped with algae bio-reactors to capture the CO_2 emissions.

Fifty million acres of switchgrass (the area of South Dakota) could produce over 140 billion gallons of ethanol per year. Battery/capacitor energy densities of over 1000 watt hours per kilogram will bring electric cars able to travel 500 miles on a single charge; eliminating the need for engines or fuel.

AUTHOR'S COVER BIO

James B. Edwards. In World War II, he was a Combat Engineer in General Patton's Third Army. After graduating from the University of Virginia, he served in government and private industry in a variety of fields including the securities business and research and development. In the 1960s, he started and managed the Navy's supergun program. In the 1970s, he started the Edwards Car Company (ECCO), which designed and manufactured specialty cars. He worked for several think tanks in the Washington, DC area. He is the author of *The Great Technology Race, Hitler: Stalin's Stooge, and There is a Silver Bullet.*

ACKNOWLEDGMENTS

I want to thank the following people for their advice and assistance in putting the book together:

Lewey Gilstrap, a member of the Johns Hopkins University Practitioner Faculty in Information Technology, taught courses in artificial intelligence and information technology to graduate candidates. He is the author of the forthcoming book, *Machine Intelligence and Robotics.*

Steve Fambro, CEO, Aptera Motors, Inc., Pasadena, CA.

Felix Kramer, California Car Initiative.

Roberto Jerez, President, Velozzi, Beverly Hills, CA.

Carl Middleton, international business consultant and author of upcoming *Great Quotations That Shaped Western Civilization,* Arlington, VA.

Breinigsville, PA USA
17 March 2010
234371BV00003B/1/P

9 781593 306427